The Physiology of
YOGA

Andrew McGonigle, MD
Matthew Huy, MSc

HUMAN KINETICS

Library of Congress Cataloging-in-Publication Data

Names: McGonigle, Andrew, author. | Huy, Matthew P., 1979- author.
Title: The physiology of yoga / Andrew McGonigle, MD, Matthew Huy.
Description: Champaign, IL : Human Kinetics, [2023] | Includes
  bibliographical references and index.
Identifiers: LCCN 2021062991 (print) | LCCN 2021062992 (ebook) | ISBN
  9781492599838 (paperback) | ISBN 9781492599845 (epub) | ISBN
  9781492599869 (pdf)
Subjects: LCSH: Yoga--Physiological aspects. | BISAC: HEALTH & FITNESS /
  Yoga | BODY, MIND & SPIRIT / Mindfulness & Meditation
Classification: LCC RC1220.Y64 M34 2023 (print) | LCC RC1220.Y64 (ebook)
  | DDC 613.7/046--dc23/eng/20220307
LC record available at https://lccn.loc.gov/2021062991
LC ebook record available at https://lccn.loc.gov/2021062992

ISBN: 978-1-4925-9983-8 (print)

Senior Acquisitions Editor: Michelle Earle; **Senior Developmental Editor:** Cynthia McEntire; **Managing Editor:** Shawn Donnelly; **Copyeditor:** Janet Kiefer; **Indexer:** Rebecca L. McCorkle; **Permissions Manager:** Martha Gullo; **Senior Graphic Designer**: Joe Buck; **Cover Designer:** Keri Evans; **Cover Design Specialist:** Susan Rothermel Allen; **Illustration (cover):** Heidi Richter/© Human Kinetics; **Photographs (interior):** Shannon Cottrell/© Human Kinetics; **Photo Production Specialist:** Amy M. Rose; **Photo Production Manager:** Jason Allen; **Senior Art Manager:** Kelly Hendren; **Illustrations:** © Human Kinetics; **Printer:** Versa Press

We thank SoundBite Studios in Los Angeles, California, for assistance in providing the location for the photo shoot for this book.

Human Kinetics
1607 N. Market Street
Champaign, IL 61820
USA

United States and International
Website: **US.HumanKinetics.com**
Email: info@hkusa.com
Phone: 1-800-747-4457

Canada
Website: **Canada.HumanKinetics.com**
Email: info@hkcanada.com

E8144

Tell us what you think!
Human Kinetics would love to hear what we
can do to improve the customer experience.
Use this QR code to take our brief survey.

The Physiology of
YOGA

Contents

Foreword

Andrew and Matt artfully revere yoga and its tradition while shining a critical spotlight on the claims teachers often make. They have identified a need within the framework of anatomy education for yoga teachers and produced this book as an exemplary way to inject critical thinking into these programs. *The Physiology of Yoga* balances research, empiricism, anecdote, humor, and passion, making it a welcome addition to any yoga teacher's library.

Yoga schools providing education at any teaching level can use this book as part of their curricula. The anatomy and physiology terms are presented in a manner appealing to yoga teachers—not too academic or medicalized yet not too rudimentary. The book attracts an enthusiastic reader who cares about a sustainable yoga practice and wants to do right by their students when stating the benefits (or risks) of the practice.

In my years as an anatomy and biomechanics educator for yoga teachers, I have found our community to be quite insular. We tend to learn from each other, share within our own spheres, and trust what our teachers say. While this makes for an impassioned community, it also supports the flow of misinformation. Fortunately, social media is beginning to change that, as is the publication of many books on the subject written by authors like myself and the authors here, who have stepped outside of the container and continued their education in other fields of study. The academic rigor of graduate level work of such writers provides the perfect catalyst to shift the narratives and allow for more uncertainty among yoga teachers. Paradoxically, being less certain creates more expertise and trust among the leaders in yoga education. As it does in science.

The Physiology of Yoga addresses the major systems of the body and the common ailments within them yoga promises to fix. The text is extremely modern, as it captures a wide range of popular opinions within the mainstream cultures of fitness and wellness. The authors seamlessly refer to research when debunking while simultaneously hinting, "Hey, if it works, it works, and if you feel better, you feel better," a sentiment often disregarded by the Western medical authority. Andrew and Matt kindly remind you that a singular anecdote or case study is not enough evidence to make powerful claims, while also appealing to your values with their own personal stories and opportunities to "try it yourself." Among all this opportunity to witness, and perhaps challenge, the reader's bias, special attention is still paid to the yoga teacher's scope of practice, which is often forgotten when therapeutic effects of yoga enter the conversation.

The final chapter includes a variety of practices for different levels of exertion, presented with surprisingly classic cueing. What I appreciate about this is the authors demonstrate it's okay to suggest when to inhale or how to align your feet. Often, recognizing that alignment rules aren't universal creates confusion and frustration because teachers now don't know what to say. The instructions within provide a non-judgmental and curious way to approach the poses without destroying the script. *The Physiology of Yoga* has my seal of approval.

Jules Mitchell, MS, LMT, RYT
Yoga Educator
Research and Adjunct Faculty at Arizona State University
Author of *Yoga Biomechanics: Stretching Redefined*
www.JulesMitchell.com

Introduction

We (the two authors of this book) have been studying anatomy and physiology in some shape or form for more than 20 years, and we still regularly find ourselves in awe of how incredible the human body is. This book is the perfect opportunity for us to share this passion and relate the science to yoga.

In recent years, we have discovered a joy in examining the many "old wives' tales" that tend to spread like wildfire through the yoga community. Does Headstand (Sirsasana) stimulate the pineal gland? Can yoga help to manage anxiety? Do twists detoxify the liver? While not everything has to be evidence-based or quantified, we believe in the importance of trying to separate theory from fact. But our ultimate aim is to inspire people to appreciate their body and love their yoga practice. Yoga and movement, after all, offer countless benefits. Our hope, in writing this book, is to inspire yoga and movement teachers to have more confidence and less fear in working with people's bodies and for practitioners to have the knowledge to help themselves in their yoga practice. Through yoga, we benefit from so much more than just a good stretch; we come to know ourselves better. We hope you learn something from this book that helps you understand yourself a little better.

Before exploring how yoga affects the physiology of each system, we must first define what we mean by yoga and physiology—and explore how everyone might benefit from critical thinking.

WHAT IS YOGA?

Yoga is a philosophical practice or discipline originating from South Asia. The word *yoga* is derived from the Sanskrit root *yuj*, meaning to join, yoke, or unite. Per yogic scriptures, the practice of yoga reminds us that not only are our mind and body connected, but our individual consciousness is inherently linked with that of the universal consciousness. The eight limbs of yoga described in Patanjali's *Yoga Sutra* are ethical principles regarding our relationship to others and the world around us (*yama*), internal disciplines (*niyama*), the physical practices of posture (*asana*), breathing exercises (*pranayama*), withdrawal of the senses (*pratyahara*), concentration (*dharana*), meditation (*dhyana*), and unutterable joy (*samadhi*). The eight limbs in themselves are not yoga but ancillary practices in support of yoga. So, while we will focus more on the physical aspects of yoga in this book, we recognize that yoga is so much more than this.

WHAT IS PHYSIOLOGY?

Physiology can essentially be thought of as the science of life. It is the branch of biology that aims to understand the mechanisms of living things, from the basis of cell function to the integrated behavior of the whole body and the influence of the external environment. The field of physiology is constantly evolving as research advances our understanding of the detailed mechanisms that control and regulate the behavior of living things. Research is also crucial in helping us to determine the cause of disease and develop new treatments and guidelines for maintaining our health.

CRITICAL THINKING

In a world of uncertainty, it can be tempting to wish for definitive answers: the one right way to practice yoga, the one right diet to follow, the one right way to live. However, particularly in a time when anyone can publish anything, it is easy to find contradictory answers to any question. This section explores the topic of critical thinking with tips on how to navigate all the conflicting views and opinions that are shared about many topics in the yoga world.

We probably like to believe that we think logically, reasonably, and without bias, but that is not always so. According to philosopher-educators Richard Paul and Linda Elder, two important figures in the development of critical thinking, "Much of our thinking, left to itself, is biased, distorted, partial, uninformed or down-right prejudiced. Yet the quality of our life and that of what we produce, make, or build depends precisely on the quality of our thought" (Paul and Elder 2019, p. 2). Thinking logically, reasonably, and without bias must be systematically cultivated, they argue.

Paul and Elder (2019) define critical thinking as "the art of analyzing and evaluating thinking with a view to improving it" (p. 2). Critical thinking is based on intellectual values that most of us would probably like to cultivate: clarity, accuracy, precision, consistency, relevance, sound evidence, good reasons, depth, breadth, and fairness. Critical thinking requires asking ourselves questions such as: *What are my beliefs on a topic? What information am I using to come to my conclusion? What assumptions have led me to this conclusion? From what point of view am I looking at this issue? Does the evidence support my conclusion? What evidence against my hypothesis exists?*

That last question is a particularly important one. We are all at times guilty of confirmation bias, a cognitive prejudice where we only seek out or listen to evidence that supports our beliefs. However, critical thinking asks us to seek out evidence against our beliefs, against our hypotheses.

Critical thinking can help us sift through all the sometimes contradictory information you might find on any given topic. One month, a study might come out showing that wine, in moderation, is good for you. The next month, a study might show that any alcohol consumption is harmful. Within the yoga community, you might hear one teacher say that sitting bones must remain on the floor during a seated forward fold, while another might encourage everyone to pull their sitting bones back. See figure 1 for a guide to using critical thinking when reading an article about a scientific finding.

EVIDENCE AND RESEARCH

Evidence is the available body of facts or information indicating whether a belief or proposition is true or valid. While most people in an asana class are probably not thinking about evidence while flowing through Sun Salutations, many yoga teachers make claims that lack any real evidence regarding the physiological workings of yoga. These claims are also often shared on yoga teacher trainings, on wellness blogs, and in books on yoga. Consider the claim that Shoulder Stand (Sarvangasana) stimulates the thyroid gland or that twists cleanse the liver. Both of these claims are based on speculation, lack evidence, and do not have sound physiological reasoning behind them.

Tip 1: Look beyond the headline. Headlines are written to catch your attention, not to tell the whole story. Often, the headline makes the finding look like more of a breakthrough than it really is.

Tip 3: How was the study conducted? Was it carried out on humans, animals, or isolated cells or tissues in a laboratory? Studies in animals and cells play a vital role in developing scientific understanding but we must not assume that their findings automatically apply to humans in real life.

The New Yoga Times

Bananas Cure Cancer

Scientists have discovered that a magic ingredient in bananas could be key to preventing cancer.

In an experiment, researchers studied what happened to cancer cells in mice when they came into contact with an extract from bananas.

They found that prostate and breast cancer cells grew more slowly and survived for less time after being treated with the banana extract.

The research was supported by the Association of Banana Manufacturers.

Tip 2: Be wary of anything claiming to be a "magic ingredient" or something similar. Most, if not all, diseases are complex with many factors and influences. Even psychological concepts such as happiness are equally complex and not down to just one factor. It is very unlikely that one food will prevent cancer or guarantee lifelong happiness.

Tip 4: Who funded the research? If funded by a manufacturer or someone with a vested interest, be particularly cautious about findings in a case of a conflict of interest.

Tip 5: How do the findings of this study relate to previous research? If the article does not place the results in the context of other studies, it is difficult to know how significant the new findings are.

FIGURE 1 Important questions to ask when looking at an article about a scientific finding.

Yet, because they are uttered so often, many yogis and teachers begin to consider them fact. But hearing something many times does not qualify as evidence. As philosopher-academic Bertrand Russell wrote, "The fact that an opinion has been widely held is no evidence whatever that it is not utterly absurd" (Russell 1929, p. 58).

Russell also used the analogy of a teapot to illustrate that the burden of proof lies upon a person making unfalsifiable claims, rather than upon others to disprove the unfalsifiable claims. He wrote that if he were to assert, with no proof, that a teapot, too small to be seen by telescopes, orbits the sun in space between Earth and Mars, he could not expect anyone to believe him solely because his assertion could not be proven wrong. It is the same with claims made about yoga.

So, evidence is simply knowledge discovered through observation and experimentation, and the best conclusions are drawn from research that has been conducted in a systematic and rigorous way—or, in other words, a scientific way. It is one thing to claim that yoga cures cancer. It is quite another to lead a thorough investigation into the effects of yoga on cancer and cancer patients. This book looks at popular claims about yoga, including the claim that yoga is good for cancer patients, and it provides the latest research to determine whether such claims are valid.

HIERARCHY OF EVIDENCE

Scientists use a hierarchy of evidence (figure 2) to help assess the quality of research. This hierarchy of evidence is often described as a pyramid, providing a visual representation of both the quality of evidence and the amount of evidence available (Sackett et al. 2000). At the top of the pyramid are systematic reviews, meaning they are both the highest level of evidence and the least common. As you descend the pyramid, the amount of evidence increases as the quality of the evidence decreases. Risk of bias also increases as you descend the pyramid.

1. *Systematic Review.* This is a review of the evidence on a clearly formulated question that uses systematic and explicit methods to identify, select, and critically appraise relevant primary research. Systematic reviews often, but not always, contain a meta-analysis of numerical data from the included studies. The methods used in a review must be reproducible and transparent.

2. *Randomized Controlled Trials.* These are studies with a randomized group of participants in an experimental group and a control group. These groups are followed up for the variables or outcomes of interest. Control groups often take a placebo to give the analysis a comparison point.

3. *Cohort Study.* Two groups (or cohorts) of participants, one which has received the exposure of interest and one which has not, are followed for the outcome of interest.

4. *Case-Controlled Study.* This is an observational study in which two existing groups differing in outcome are identified and compared on the basis of some supposed causal attribute.

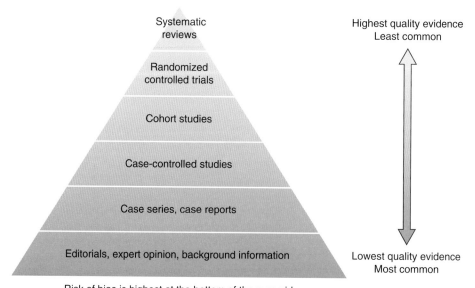

Risk of bias is highest at the bottom of the pyramid

FIGURE 2   Hierarchy of evidence as a pyramid.

5. *Case Series, Case Reports.* A case report is a detailed report of the symptoms, signs, diagnosis, treatment, and follow-up of an individual. Case reports may contain a demographic profile of the patient but usually describe an unusual or novel occurrence. A case series is a type of study that tracks multiple individuals with a known exposure, such as people who have received a similar treatment, or the researchers examine their medical records for exposure and outcome.

6. *Editorials, Expert Opinion, Background Information.* This level of evidence includes articles in newspapers or other publications presenting the opinion of the author, or background texts, including textbooks, which provide a broad overview of a topic with a selected review of the scientific literature.

While this pyramid offers a useful guide for assessing the quality of evidence, it has its limitations. Randomized control trials (RCTs) are given the highest status of all experiments because they are, by design, less biased and have the least risk of systematic errors. RCTs work well in drug studies because it is easy to disguise the placebo as an active drug. RCTs do not, however, work so well for interventions like yoga. You cannot hide from a study participant the fact that they are doing yoga, and it is difficult to use a placebo in lieu of yoga. The pyramid also does not account for qualitative research, which is not experimental in design but instead uses nonnumerical data (e.g., text, video, or audio) to understand concepts, opinions, or experiences. Qualitative research might explore, for example, how yoga makes someone feel or how yoga affects their quality of life—and the pyramid does not neatly cover this type of research.

In this book, we look mainly at systematic reviews, which are considered a higher level of research. While we searched many scientific databases for reviews on the physiological effects of yoga, this book is narrative in nature and thus is a selected review of the scientific literature. Just as with any review, this book has its limitations.

Good writing should include references, which is why you will find in-text citations with author names and years throughout the book and a reference list at the end of the book. With scientific writing, the reader should be able to track down the same information as the author.

One final note: Throughout this book, you will see the term *et al.*, which is short for the Latin *et alia*, meaning "and others." It is standard practice in scientific writing to use *et al.* instead of including all author names.

Now that we have defined *yoga*, *physiology*, and *critical thinking*, we can begin our journey into the body, starting with the musculoskeletal system.

Chapter 1

MUSCULOSKELETAL SYSTEM

The musculoskeletal system, once known as the *activity system*, is what gives us the ability to freely move our body while also providing support and stability. The musculoskeletal system is composed of the muscles and the skeleton but also the cartilage, tendons, ligaments, joints, and other connective tissue that support and bind tissues and organs together.

Many people come to yoga for the benefit of the musculoskeletal system. Perhaps they have a general feeling of tightness, or a physical therapist has recommended yoga. Many fanciful claims are made about how yoga—and stretching in general—can improve the health of this system, and the topic of fascia has become a popular topic within the yoga community. It can sometimes be difficult to know what is true and what is not. Does yin yoga work on the fascia of the body? Can yoga wear down musculoskeletal tissues? What is the best type of stretching? What can we do to improve resilience and decrease injury? How yoga actually affects the musculoskeletal system may surprise you. A little background on the system will be useful.

BONE

While bone may seem solid and inert like the plastic skeletons in a doctor's office or the white remains found in a museum, the bones in your body are very much alive, playing many life-essential roles and even adapting to the demands placed on them.

Structure of Bone

While bones have different shapes, they all have the same basic structure (figure 1.1). The most superficial layer of the bone is a thin covering of connective tissue called *periosteum*. This periosteum contains nerves that sense pressure. This is how you know you have banged your shin bone into something. Blood enters bones through blood vessels that flow through the periosteum.

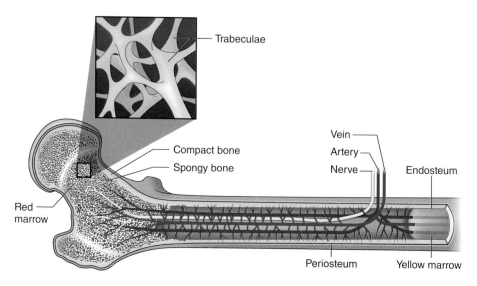

FIGURE 1.1 Layers of bone.

Beneath the periosteum is a hard layer called *cortical bone*, which is largely composed of proteins like collagen and hydroxyapatite, a naturally occurring mineral with calcium as its main constituent. Hydroxyapatite, which makes up 70 percent of the bone's weight, is responsible for making bone strong and dense. Bones also store 98 percent of the body's calcium (Institute of Medicine 2011), which is an essential component of any muscular contraction. Bone releases calcium as needed, then reabsorbs it when not needed.

The next layer, called *trabecular bone*, is softer and less dense but still makes a significant contribution to the strength of the bone. A reduction in the quality or volume of trabecular bone increases the risk of fractures (breaks). Finally, bone marrow fills the spaces in the trabecula. Red bone marrow contains specialized cells that produce red blood cells (which carry oxygen around the body), white blood cells (important in fighting infection), and platelets (important in clotting blood), while yellow marrow is important for fat storage. Blood vessels supply blood to the bone and receive the newly formed cells from the marrow.

Being rigid structures, bones maintain the form of the body and protect internal organs. But they also create the framework for movement. All voluntary movement, including all movement performed in a yoga practice, happens at joints, which is where two bones articulate. And the loading that occurs during asana practice and other weight-bearing activities is very important to the health of bone.

How Bones Adapt

Without your realizing it, your bones are adapting every day. While you might not give much thought to gravity, which is pulling you toward the center of the earth at a rate of 9.8 meters per second squared, your skeleton is constantly adapting to it. In the absence of gravity, as happens with astronauts in space flight, significant losses in bone mass occur. In fact, astronauts lose an average of 1 to 2 percent bone mass per month in space in a phenomenon known as *spaceflight osteopenia* (Kelly and Lazarus Dean 2017, p. 174; NASA 2001). Most of the loss occurs in the lower limbs and lumbar spine, with the proximal part of the femur losing roughly 10 percent of its bone density for every six months in space, even though the astronauts exercise 2.5 hours per day, six days a week, using springs and vacuum canisters for resistance (NASA 2001).

To maintain bone density and strength, our body requires an adequate supply of calcium and other minerals as well as vitamin D, and the endocrine system must also produce the proper amounts of several hormones, such as parathyroid hormone, growth hormone, calcitonin, estrogen, and testosterone. (More information on hormones can be found in chapter 6 on the endocrine system.) Another requirement is adequate mechanical loading to induce remodeling.

Well before the advent of space flight, Julius Wolff first proposed in 1892 that bone adapts its architecture to the stresses put on it, a concept that came to be known as *Wolff's law* (Wolff [1892] 1986). In 1964, Harold Frost refined this observation to reflect the knowledge that bones are not straight but slightly curved structures, and the mechanostat model was born (Frost 1964). As loads are applied to the body, this mechanical stimulus is converted into electrochemical activity in a process known as *mechanotransduction*. These signals are sent to the central nervous system, which responds by instructing the bone to build a stronger and denser framework to support the new demands. Osteoblasts are bone-making cells that initially live on the outside of the bone, turn into osteocytes, and become embedded within the bone, causing new bone to be laid down (Turner and Pavalko 1998). Meanwhile, osteoclasts break down older, damaged, or unhealthy bone tissue so the materials can be reabsorbed for new bone (Robling, Castillo, and Turner 2006).

In short, our nervous system senses how much bone strength is required and adapts accordingly. This process is happening so frequently that every bone in the body is completely reformed about every 10 years (Manolagas 2000). With physical activity and adequate amounts of hormones, vitamins, and minerals, trabecular bone develops into a complex lattice structure that is lightweight yet strong. In addition to the external force provided by gravity, the internal force provided by muscular contraction can also provide enough stimulus to elicit bone adaptation. As muscles contract, they pull on bones, and that tugging can create enough stimulus to increase the strength of bone (Russo 2009). In the section about osteoporosis on page 24, we look at whether yoga provides enough stimulus to elicit strengthening of bone.

JOINTS

Coming from the Old French for joined, a joint is the point where two bones come together. While a bone is a solid structure, a joint is, in a way, a nonstructure. It is the space between two bones; it is a relationship between two structures. Fibrous joints permit very little movement in order to protect the organs behind them. One such example is the sutures between the bones of the skull, which hardly move (except during birth) in order to protect the brain. Synovial joints, in contrast, are self-lubricating and provide near-frictionless movement and the bearing of heavy loads. Synovial joints are the ones we move a lot in yoga asana and include the elbow, shoulder, hip, and knee.

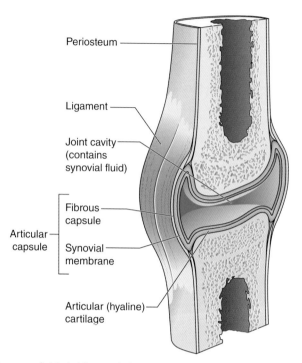

Periosteum

Ligament

Joint cavity (contains synovial fluid)

Fibrous capsule

Articular capsule

Synovial membrane

Articular (hyaline) cartilage

FIGURE 1.2 A synovial joint has a joint capsule that encloses the joint, articular cartilage that protects the ends of bones and provides a smooth surface, and synovial membranes that release synovial fluid, a lubricant for the joint.

All synovial joints have a joint capsule (formed of fibrous connective tissue, forming the container of the joint), a synovial membrane (which releases synovial fluid, the lubricant of the joint), and articular cartilage (which protects the ends of bones) (figure 1.2). The combination of smooth articular cartilage and synovial fluid makes for a near-frictionless surface. During movement, the thickness of the synovial fluid can change, as happens with non-Newtonian fluids, which helps to make a joint more stable as it moves.

MUSCLES

While all skeletal movement happens at joints, muscles are the motor behind the movement. Muscles—coming from the Latin word for mouse, perhaps because of how they look as they glide under the skin—generate force, which allows us to maintain and change posture, lift objects, move our bodies through space, and involuntarily pump blood through our arteries and move food through our digestive tracts.

There are three types of muscle: cardiac, smooth, and skeletal. Cardiac muscle forms the walls of the heart, pumping blood every minute of every day. Smooth muscle is found in a variety of areas and serves functions such as contracting the arteries to control blood pressure, raising the small hairs on your arm, and moving fluids through organs by applying pressure to them. Smooth muscle does not contract or release as quickly as skeletal or cardiac muscle, but it is much more useful for providing consistent, elastic tension. Both cardiac and smooth muscle contract without conscious thought and thus are termed *involuntary*. Skeletal muscle (also called *striated muscle*), in contrast, engages upon command and thus is termed *voluntary*. Skeletal muscles allow us to move our bodies in the innumerable ways we do in a yoga class.

How Muscles Contract

Muscle tissue needs a nerve impulse in order to move. Cardiac and smooth muscle contractions are stimulated by internal pacemaker cells. In the heart, this pacemaker is the sinoatrial node and, in the gut, the enteric nervous system. With skeletal muscle, the impulses come from the central nervous system (CNS), which is the brain and spinal cord, via motor neurons.

While these three muscle types have significant differences, they all move following the same principle. Muscle cells contain protein filaments of actin and myosin that slide past one another, generating force that can change the length and the shape of the cell. The proteins actin and myosin form crossbridges and slide along each other to create contraction, following the sliding filament theory, which was first proposed in 1954 (Krans 2010). Each individual muscle cell (also called a *muscle fiber*) contains long units of myofibrils, and each myofibril is a chain of tens of thousands of sarcomeres (figure 1.3). These long chains shorten together, shortening the muscle fiber and creating a contraction. Although the evidence behind the sliding filament theory is quite solid, muscle contraction is still not completely understood. For example, titin is an unusually long and spring-like protein spanning many sarcomeres and appears to bind to actin, but it is not well understood (Krans 2010). Researchers are currently exploring the role of titin, which will eventually help us to better understand the mechanisms of muscular contraction.

Types of Contraction

Muscle contractions are described using two variables: force and length. When muscle tension, which simply means force exerted by the muscle, changes without any corresponding changes in muscle length, the muscle contraction is isometric. The prefix *iso-* means same, and *metric* refers to length, so *isometric* means same length. When the muscle length changes while muscle tension remains the same, the muscle contraction is isotonic, meaning same tension. Within the isotonic category, a muscle can either shorten to produce a concentric contraction or lengthen to produce an eccentric contraction. The word *concentric* can be translated as toward the center, while *eccentric* can be translated as away from the center. Some people struggle with the term *eccentric contraction* because the word *contraction* has a suggestion of shortening, whereas eccentric contraction is a lengthening movement. Some find it more helpful to use the term *effort*, thus referring to these movements as *eccentric effort*, *concentric effort*, or even *isometric effort* (figure 1.4).

Analogizing these movement descriptors to a car can be helpful. Imagine you are stopped in a car on a steep hill. As you release the brakes, one of four things could happen:

1. You could free-fall backward down the hill if you were in neutral. This would be the same as simply dropping your arm from an elevated height and not controlling the downward movement.

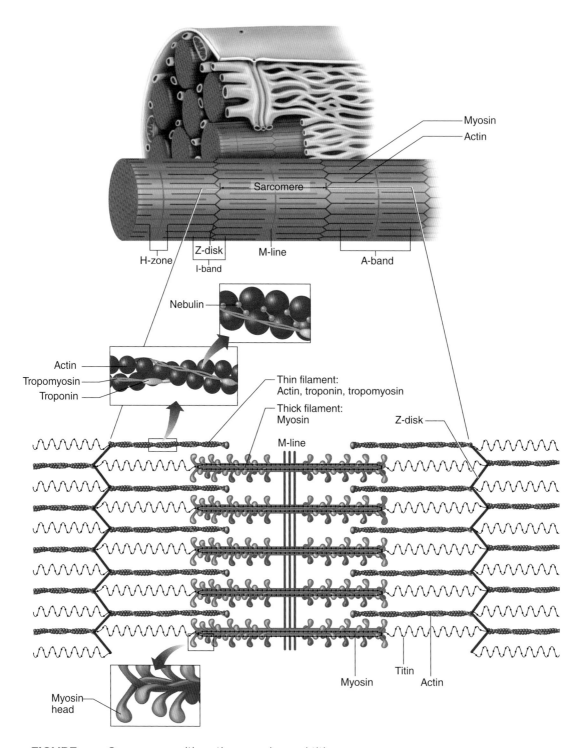

FIGURE 1.3 Sarcomere with actin, myosin, and titin.

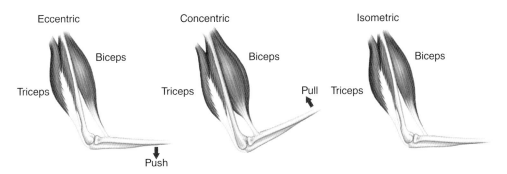

FIGURE 1.4 Types of contraction: eccentric, concentric, and isometric.

2. You could apply a small amount of gas so that you slow the backward descent. In this instance, you are using some energy to control the downward movement. This would be an eccentric effort of muscle.

3. You could use more gas until the backward movement stopped and you remained on the hill in the spot where you released the brake. This would be an isometric effort of muscle.

4. Finally, you could use even more gas and start to move uphill. This would be a concentric effort of muscle.

As you might surmise, yoga involves a lot of isometric effort. Whenever you hold an active asana, your muscles must engage isometrically. Maintaining upright posture throughout the day is also an isometric effort. Transitioning in and out of poses, however, requires concentric and eccentric efforts of various muscles.

Eccentric effort is of particular interest to researchers. While the sliding filament theory neatly explains concentric contractions, current muscle models have difficulty explaining the increased force and reduced energy cost of eccentric contractions (Hessel 2018). Even calling muscular effort a contraction suggests how researchers have historically focused on explaining the phenomenon of muscles shortening to generate force. Some researchers now believe titin—the long, spring-like protein mentioned previously—might play an important elastic role (Hessel 2018).

CARTILAGE

Cartilage is a smooth and resilient type of connective tissue with a firm, gel-like constitution. The matrix of cartilage is made up of glycosaminoglycans, proteoglycans, collagen fibers, and, sometimes, elastin. Our bodies have three major types of cartilage: hyaline cartilage, elastic cartilage, and fibrocartilage. Hyaline cartilage, the most common in the body, provides support that is stiff but somewhat flexible. Examples include part of the nasal septum and the ends of the ribs where they meet the breastbone—perfect for being rigid enough to help protect the heart and lungs inside but flexible enough to move and expand with the breath. Another example of hyaline cartilage is at the ends of bones, where it is called *articular cartilage*. With its slick and smooth surface, glass-like articular cartilage is an ideal covering of the bone. Cartilage on cartilage lubricated by synovial fluid is incredibly slippery—significantly

Use Eccentric Training to Your Advantage

We can bear greater loads eccentrically than isometrically—and we can bear greater loads isometrically than concentrically. While this might sound complex, we can all relate to this principle in real life.

Imagine performing a biceps curl with a dumbbell. Let's say you can control the downward movement of a 40-pound (18 kg) dumbbell from next to your shoulder down to your hip (extension of the elbow and an eccentric effort). But if someone asked you to stop midway through that movement (an isometric effort), you might not be able to. So you reduce the weight to a 30-pound (14 kg) dumbbell. You can control the downward movement of this lighter dumbbell, and now you can hold the dumbbell in place midway as well. But if someone asked you to lift the dumbbell, the load might be too great for you. You change to a 20-pound (9 kg) dumbbell. You can control it eccentrically and isometrically. Then when asked to lift it (a concentric effort), you can.

This concept can also be applied to body-weight movements found in yoga. Imagine coming into Headstand (Sirsasana) or Handstand (Adho Mukha Vrksasana) through a pike press, which is a concentric movement (figure 1.5). If you are not able to do this, one way of progressing toward it is to build strength and competence with eccentric movements. To do this, come into Headstand or Handstand and attempt to slowly lower your feet to the floor with the legs straight. You might find a breaking point when the legs simply collapse to the floor. Try to isometrically hold right before you get to that breaking point. After building strength over a matter of weeks with these eccentric and isometric contractions, you will probably find that you are closer to achieving a controlled pike press (Vogt and Hoppeler 2014). Because we can bear greater loads eccentrically than we can concentrically, eccentric training can enhance maximal muscle strength and power even faster than solely concentric training. Plus, it can further optimize maximal tension at a greater degree of extension (Vogt and Hoppeler 2014). So, if you have ever wanted to do a pull-up but do not know where to begin, start by jumping up to the bar and lowering yourself down as slowly as possible.

a   b

FIGURE 1.5 Eccentric training. If you want to learn to do a pike press to Handstand, *(a)* come into your Handstand however you normally do, then *(b)* lower yourself as slowly as possible. Aim to isometrically hold at various points during the descent until you are strong enough to concentrically do a pike press from the floor. The same can be applied to Headstand.

more slippery than an ice skate gliding on ice (Guilak 2005; Mow and Ateshian 1997). This helps to reduce friction during joint movement.

Elastic cartilage has many elastic fibers in addition to collagen, making the matrix much more elastic than hyaline cartilage. Elastic cartilage provides support but can tolerate deformation without damage and can return to its original shape. It is found in the external ear, epiglottis, and larynx (voice box).

Fibrocartilage has an abundance of thick collagen fibers and compounds (glycosaminoglycans) that absorb water, making it tough and deformable, suitable for its role in intervertebral discs and insertions of tendons. It resists compression, prevents bone-to-bone contact, and limits excessive movement. Fibrocartilage is found in the menisci of the knee, between the pubic bones of the pelvis, and between the vertebrae of the spine.

Muscle tissue has a very rich supply of blood and can thus easily repair itself from damage, whether that damage is through the controlled stress of exercise or through injury. Cartilage, however, does not have a blood supply, and so it is understood that cartilage is very poor at remodeling or regenerating itself once damaged. This inability of cartilage to repair well is believed to contribute to osteoarthritis—inflammation in the joint that causes pain and stiffness.

However, recent research has shown that human cartilage might, to some degree, be able to renew itself despite its lack of blood supply. Looking at articular cartilage that had been removed from patients having joint surgery, Hsueh and colleagues (2019) found that human cartilage uses microRNA in a complex molecular process similar to the one that allows a salamander to grow a new limb. This research showed for the first time that some articular cartilage has the potential to repair itself, but this capability exists in a gradient, greatest at the ankle, less in the knee, and lowest in the hip. The researchers also posited that this finding might partly explain why osteoarthritis is common in the knees and hips but not the ankles. This new insight, the researchers hope, might contribute to new treatments for arthritis.

Though this one study has revealed some novel findings, cartilage repair is a complex issue, and we do not have all the answers. For example, while it was previously believed that the outer third of the knee's meniscus—crescent-shaped cartilage that helps with shock absorption and weight distribution—had the potential to repair itself, recent evidence found no detectable remodeling using radiographic imaging (Våben et al. 2020).

Even though there are still many questions around the capability of cartilage to repair itself, here are two things we know:

1. Movement is nourishing. Because cartilage has little to no blood supply, particularly articular cartilage found within the joint capsule, cartilage cells (called *chondrocytes*) rely on the slow process of diffusion to move nutrients into the joint capsule and waste materials out. That process is dramatically sped up with joint movement. This might be one of the reasons that most people with arthritis say they feel better when they move.

2. Cartilage needs mechanical stimulation. What little remodeling cartilage might be able to manage occurs as a result of mechanical stimulation of the tissue. Loading and unloading of the cartilage that happens through yoga, exercise, walking, and running can help to keep cartilage healthy.

TENDONS, LIGAMENTS, AND APONEUROSES

While tendons and ligaments serve different purposes, they are mostly the same composition and thus often described together. In short, tendons connect muscle to bone, transmitting the force generated by the muscle to move the skeleton at a joint. Ligaments connect bone to bone, providing stability and limiting movement at a joint. Both are composed of collagen-rich, dense fibrous connective tissue.

Tendons—coming from the Latin *tendere*, which means to stretch or to draw tight—are tensioned as muscle contracts, thus transmitting large mechanical forces from the muscle to the bone, eliciting movement. Each muscle has two tendons—one proximal and one distal. The point at which the tendon becomes the muscle is known as the *musculotendinous junction* (or *myotendinous junction*), and the point at which it attaches to the bone is known as the *osteotendinous junction*. The reality is, however, that the connective tissue that is the tendon becomes the connective tissue that envelops and even inserts into the muscle tissue. At the osteotendinous junction, the connective tissue that is the tendon becomes the connective tissue that covers the bone, which is known as *periosteum*.

One way that our body adapts to exercise is by increasing tendon stiffness (Reeves 2006). Though stiffening a tendon might not at first sound like a beneficial adaptation, consider the following: imagine a horse drawing a cart, which is akin to muscle pulling bone. If the horse were connected to the cart via a band of stretchy rubber, much of the horse's effort would be lost in the stretching of the band. A better device would be a length of stiff rope that would stretch less than the rubber, and thus more force would be transferred efficiently. By becoming stiffer through adaptation, tendons become more efficient at transmitting force to the bone to allow us to move our bodies in the way we want.

Similar to tendons are aponeuroses. An aponeurosis, a type of deep fascia that is structurally similar to tendons and ligaments, is a sheet of dense fibrous tissue that anchors a muscle or connects it with the part that the muscle moves. It is sometimes described as a long, flat tendon, as opposed to a cord-like tendon. However, it may connect muscles to other muscles, as opposed to tendons, which connect muscle to bone. The primary regions with thick aponeuroses are in the ventral abdominal region, the dorsal lumbar region, and the palmar (palms) and plantar (soles) regions.

FASCIA

Fascia has become a buzzword in the yoga community. Here are some of the many claims you might hear about fascia within the yoga or massage community:

- Emotions are stored in the fascia.
- Tightness is in the fascia, not the muscles.
- Fascia gets stuck if we don't move and stretch.
- Yin yoga targets the fascia.
- Massage and foam rolling help break down fascial adhesions.

At one time, little regard was given to fascia. Andrew (one of the authors of this book) recalls that when he was observing dissections of cadavers during his medical training, the dissectors cut away the white stuff around organs and muscles and threw it in the waste bucket. When asked what they were throwing out, they replied, "That's just fascia."

Today, though, researchers give more regard to fascia—so much so that the Fascia Research Congress was established in 2007 with a meeting occurring every three years to present the latest findings about this previously discarded tissue. Though fascia research is published at a greater rate today than years past, many questions still remain unanswered, and the relevance of this tissue to yoga is not absolutely clear.

The first question is the definition of fascia. There is not one single definition of fascia that is uniformly accepted across the medical and scientific community. Traditionally, fascia, as laid out in *Gray's Anatomy*, was described as "masses of connective tissue large enough to be visible with the unaided eye" (Standring 2004, p. 42). The Fascia Research Society, having consulted many experts on the topic, defines fascia as "a sheath, a sheet, or any other dissectible aggregations of connective tissue that forms beneath the skin to attach, enclose, and separate muscles and other internal organs" and goes on to say:

> *The fascial system consists of the three-dimensional continuum of soft, collagen containing, loose and dense fibrous connective tissues that permeate the body . . . The fascial system surrounds, interweaves between, and interpenetrates all organs, muscles, bones, and nerve fibers, endowing the body with a functional structure, and providing an environment that enables all body systems to operate in an integrated manner. (Adstrum et al. 2017, p. 175)*

In simpler terms, fascia is commonly described as sheets and webs of fibrous connective tissue found everywhere in the body—and, in even simpler terms, "that white stuff that you find in and around your chicken." For anyone who has eaten chicken, think of the membrane that attaches the skin to the meat, or to the stringy connections between muscle parts. This cobweb-like material is one type of fascia and runs through the structures of the body, including the muscles.

The Fascia Research Society's definition explicitly includes many types of connective tissue: tendons, ligaments, neuronal coverings, sheaths around bones, and even fat tissue. And there is a good reason for thinking in this way. Anatomy books usually describe fascia as distinct and discrete regions of connective tissue. This approach is absolutely necessary to medicine and science so that we can accurately talk about different structures at different parts of the body. But the body is not simply an assembly of many discrete parts, like a smartphone, that can be disassembled and reassembled. Fascia is, instead, an interconnected web that binds everything together (figure 1.6), and so, fascia researchers argue, it should be considered as one whole system rather than individual parts.

Consider, for example, the fascia of the thigh. Anatomists label the fascia lata (a sheath of fascia covering all the muscles of the thigh like a stocking) as distinct from the iliotibial tract (the band of fascia that spans the length of the outer thigh from the pelvis to the top of the outer shin, the tibia). But in reality, there is no separation between the fascia lata and the iliotibial tract. The iliotibial tract is simply the densest portion of

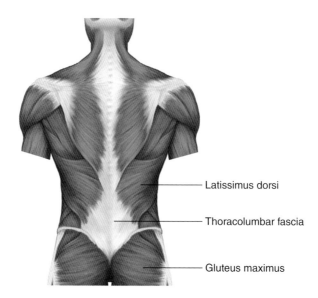

Latissimus dorsi

Thoracolumbar fascia

Gluteus maximus

FIGURE 1.6   Though the thoracolumbar fascia is often described as a discrete structure, as shown in this image, it is really just a denser part of the fascia that covers all of the muscles around it and beyond. The fascial system is indeed an interconnected system.

the fascia lata. Fascia experts are instead suggesting that, although individually labeling regions of fascia is useful, it is also important to view all fascia as one whole system, the fascial system. This idea of a body-wide fascial net becomes more apparent during the process of viewing human cadavers. When exploring a cadaver—as opposed to looking at a textbook—one can more clearly see the interconnectedness of the fascia. This is surely why fascia researcher-educators such as Gil Hedley, Tom Myers, and Robert Schleip teach about fascia through dissection.

Through dissection, Tom Myers (2014) came up with his famous *Anatomy Trains*, a book and fascia-mapping system. He notes the disadvantage of studying anatomy from the perspective of isolated muscles:

> *The common method of defining muscle action consists of isolating a single muscle on the skeleton, and determining what would happen if the two ends are approximated . . . This is a highly useful exercise, but hardly definitive, as it leaves out the effect the muscle could have on its neighbors by tightening their fascia pushing against them. It also, by cutting the fascia at either end, discounts any effect of its pull on the proximal or distal structures beyond. (Myers 2014, p. 2)*

By "turning the scalpel on its side," he was able to see long trains of fascia that function together to produce movement, which he calls *myofascial trains* (Myers 2014, p. 47). The prefix *myo-* means muscle, so *myofascia* is used to describe the nearly inseparable structure that is the muscle–fascia unit.

The Role of Fascia

Researchers have found that fascia is more than just an inert wrapper for muscles and organs. Fascia is well connected to the central nervous system with multitudes of proprioceptors (which sense body position) and nociceptors (which sense tissue damage) (Krause et al. 2016). It has also been observed that contractile cells are present in fascia, making it able to contract like smooth muscle (Schleip et al. 2005). However, that degree of contractility is such a small fraction of the contractility of muscle cells that it is not clear how significant or important this may be to the overall musculoskeletal system.

Fascia may also play a role in force transmission within the body, which might be connected to its ability to contract. A review by Krause and colleagues (2016) points to the fact that fascia can transfer tension to at least some adjacent myofascial structures. This means that when the hamstrings are stretched (tensioned), that tension can be transferred to other nearby structures, such as the gluteal muscles and the lower back as well as down to the calf muscles, seemingly along myofascial chains, giving credibility to the theory of myofascial chains in the body. While this might not sound surprising to some, this thinking provides a model very different from the traditional view that discrete muscles or muscle groups act in isolation to perform a task, such as a biceps curl.

This more holistic model of the body as a connected web of fascial chains would help explain some interesting findings such as the connection between tight calf muscles and plantar fasciitis, how strength imbalances between adductor longus and lower abdominal muscles are associated with groin pain in athletes, how ankle range of motion (ROM) seems to be affected by forward head posture, how passive hamstring stretching tends to increase cervical spine ROM, and how self-myofascial release on the plantar fascia increases sit-and-reach performance (Krause et al. 2016). These interesting findings might give more credence to the practice of yoga, which teaches that the body must be seen as a whole and not a series of disparate parts. These findings also point to the body as a tensegrity structure.

Tensegrity is a portmanteau of "tensional integrity" coined by the visionary architect, author, and designer Buckminster Fuller. It is a structural principle based on a system of isolated components under compression within a network of continuous tension, arranged in such a way that the compressed members do not touch each other. The concept is best understood through example. An unerected camping tent is a jumble of fabric, poles, and guy lines. But once the last pole is put into place, the tent becomes a stable, resilient structure able to withstand the elements due to the compression on the poles and the tension on the fabric.

When applied to biological structures, this principle is sometimes referred to as *biotensegrity*. Within the body, muscles, bones, fascia, ligaments, tendons, and other structures are made strong through the unison of tensioned and compressed parts. The muscles maintain tension through a continuous network of muscles and connective tissues, while the bones float in the connective tissue, providing discontinuous compressive support (Souza et al. 2009). Even the human spine—which might seem at first glance a stack of vertebrae resting on each other—can be described as a structure of tensegrity where the vertebrae and discs are compression members and the ligaments are tension members (Levin 2002).

Some researchers have postulated that the state of the fascia in the mid- to low back may be connected to low back pain. Langevin and colleagues (2009) found through ultrasound imaging that people with chronic low back pain had around 25 percent thicker connective tissue in the low back compared to healthy subjects. That is a substantial amount, but this research simply shows a connection, not a cause. What is not known from this study is whether thicker fascia is a contributing factor to back pain, a result of back pain, or even unrelated to back pain! Possible causes of the thicker connective tissue, the authors noted, include genetic factors, abnormal movement patterns, and chronic inflammation. So, while these findings are interesting and novel, they do not point to an obvious clinical significance of fascia.

Langevin and colleagues set out again in 2011 to explore movement of the thoracolumbar fascia in people with persistent low back pain. They found that those with pain had about 20 percent reduced shear strain—a measurement of the deformation of a structure. In simpler terms, their low back fascia was 20 percent stiffer than in people without back pain. Again, this is a novel and interesting finding, but it still does not provide a clinical significance to fascia.

It is clear that fascia plays a more important role than simply being an inert placeholder as previously thought, and better understanding the interconnectedness of the fascial system may help with future approaches to injury rehabilitation or strength and conditioning coaching. For now, however, more research is needed before we can draw strong conclusions between fascia and yoga as well as other topics like pain and overall health.

Myofascial Release

Myofascial release (MFR) has become increasingly popular and widely used since the 1990s. Although its roots can be tracked to the 1940s, the term was first coined in 1981 by three professors in a course at Michigan State University titled "Myofascial Release" (Manheim 2008).

MFR is, firstly, a treatment modality wherein the therapist applies a low-load, long-duration stretch to the soft tissues of the body, guided by feedback from the recipient's body to determine stretch direction, force, and duration. Think of it as being gently stretched by someone, perhaps with some gentle pressure from the therapist's hands along the tissues being stretched.

Some osteopathy textbooks claim that fascia can become restricted due to psychogenic disease, overuse, trauma, infectious agents, or inactivity, potentially resulting in pain, muscle tension, and diminished blood flow (DiGiovanna, Schiowitz, and Dowling 2005). MFR is claimed to help release those fascial restrictions.

As for its efficacy, reviews published in 2013 (McKenney et al.) and 2015 (Ajimsha, Al-Mudahka, and Al-Madzhar) found some encouraging results with MFR but determined that because of the low quality of the studies available, few conclusions could be drawn. A further review of MFR on chronic pain by Laimi and colleagues (2018) found that current evidence of MFR in chronic pain relies on only a few studies, that previous positive conclusions could not be confirmed, and that it is not known whether MFR is any more effective in treating chronic musculoskeletal pain than sham procedures.

Perhaps, however, MFR cannot be reliably measured through science. In an article titled "Why Myofascial Release Will Never Be Evidence-Based" for the journal *International Musculoskeletal Medicine*, Robert Kidd (2009) noted, "Myofascial release is an art form. Much depends on the innate talent and experience of the therapist" (p. 55). If an intervention relies on a therapist's innate talent and nothing else, then it cannot be easily quantified or codified and thus cannot be easily studied.

Yin Yoga Targets the Fascia

Yin yoga has steadily grown in popularity since 2011. It is commonly claimed that yin yoga works on the connective tissues of the body, particularly the fascia. Yin teachers talk about stressing the tissues of the body and distinguish it from restorative yoga. While restorative yoga is about resting in comfortable positions, yin should have some degree of discomfort or challenge to elicit change in the tissues of the body (Clark 2012). But can this slow, passive style of yoga wherein poses are often held for three to five minutes really target the fascia?

Participants in a yin class are often asked to release all muscular tension and relax into the pull of gravity, whereas participants in a vinyasa class might be asked to cocontract antagonist muscles or core-stabilizing muscles. No matter what instruction is given, a stretch is a stretch—and a stretch, as we normally define it, is a tensile force on a muscle. When a tensile force is applied to a muscle, it is also being applied to the surrounding fascia and investing muscle fibers and bundles. The muscle and fascia are so interwoven that you cannot choose which one you are stretching by contracting or not contracting certain other muscles.

In vinyasa yoga, Standing Forward Fold is called *Uttanasana*. In yin yoga, it is called *Dangling*. Biomechanically and physiologically, there is little to no difference between the two as long as the duration of the stretch (tensile force) is the same between the two poses. Additionally, given what we now know about reciprocal inhibition (discussed later in the chapter), it also appears that engaging the antagonist muscle group, the quadriceps, does not affect the stretch very much at all. A standing forward fold, by any name, is a static stretch, and all stretching is a tensile force applied to a myofascial unit. It should be noted that as tensile force is applied to tissues, they creep. *Creep* is the biomechanical term for the deformation of viscoelastic tissues. Once the tensile force is removed, tissues then recover and return to their original length, as long as they have not been elongated beyond their elastic capacity.

One study (Ryan et al. 2010) looked at creep in the muscle–tendon unit of living humans during a 30-second stretch, finding that the greatest amount of creep was measured to occur within the first 15 to 20 seconds. Beyond that, to our knowledge, no studies exist on yoga poses and creep, which means we do not know the ideal duration for stretching tissues or how long tissues take to fully recover from their creep.

So, yes, yin yoga does affect the viscoelastic fascia of our bodies but not any more than the same stretch performed in a different way (i.e., a more active way). Yin yoga, or any stretch held for three to five minutes, will affect the fascia, but the ideal frequency and duration remain a mystery.

MFR as described requires a therapist to perform the treatment on a recipient. However, self-MFR does not require a therapist. Foam rolling is probably most associated with self-MFR, but many other products and modalities—from spiky massage balls to massage sticks to percussive massage guns—also claim to release fascia. But not all of these products have been through the rigor of scientific study.

There is currently no regulation around MFR claims, so anyone with any product can claim that it elicits MFR.

A recent review by Wiewelhove and colleagues (2019) looked at all the literature available on the efficacy of self-massage products applied before exercise (as a warm-up activity) and after exercise (as a recovery strategy) on sprint, jump, and strength performance as well as flexibility and muscle pain outcomes. Twenty-one studies met their criteria—14 of which used foam rollers and 7 of which used roller massage bars or sticks. They concluded the following:

> Overall, it was determined that the effects of foam rolling on performance and recovery are rather minor and partly negligible, but can be relevant in some cases (e.g., to increase sprint performance and flexibility or to reduce muscle pain sensation). Evidence seems to justify the widespread use of foam rolling as a warm-up activity rather than a recovery tool. (p. 1)

Even if self-MFR has the potential for reducing pain sensations or improving performance, the reason is not at all clear. In 2019, Behm and Wilke looked at possible mechanisms behind any efficacy of MFR in their aptly named paper, "Do Self-Myofascial Release Devices Release Myofascia?" Examining the physiology and biomechanics of several self-massage rolling devices, they concluded that there is some evidence that rolling might affect blood flow or local hydration changes, but manual forces are not typically sufficient to change the shape of connective tissue. They also looked at research that has shown rolling to increase our tolerance to stretching and decrease pain, determining that this is likely due to counterstimulation. Counterstimulation is the process of decreasing pain signals from a site because of the introduction of a more obvious stimulation (think of rubbing your shin after bumping it or biting your lip when receiving a jab). They also suggested that increased parasympathetic activity (in other words, being more relaxed) might be an important mechanism behind rolling's observed benefits. So, they concluded, the primary mechanism behind rolling and other similar devices is not the release of myofascial restrictions, thus suggesting that the term *self-myofascial release* is a misleading one.

Myofascial release, whether administered by a therapist or by oneself, might very well have some benefits—and if you personally swear by it to keep yourself pain-free and injury-free, then that might be the only evidence you need. However, the scientifically observed benefits do not (so far) seem particularly significant. The mechanism behind any of MFR's benefits seems likely due to temporary changes to the nervous system, rather than the fascia, suggesting that the term *myofascial release* is a misnomer. Additionally, a product does not need any science behind it to be labeled a myofascial-releasing device, so caution is advised when reading the label of any such item.

THE SCIENCE OF STRETCHING AND FLEXIBILITY

With its focus on holding positions at end ranges of motion and advanced postures that look very similar to the stunts of a contortionist, an undeniable connection between yoga and flexibility exists. Yoga is widely considered a practice of flexibility, and flexibility is the number one reason people start yoga, according to a survey by Yoga Alliance (2016).

There are multiple definitions of flexibility and mobility depending on the source and the context. Some recent definitions of *flexibility* refer to a muscle's ability to lengthen passively through a range of motion, while *mobility* is sometimes used to refer to the ability of joints to move actively through their range of motion. For this book, we refer to flexibility as joint ROM and break it down between active and passive ROM. Joint ROM can be measured using a goniometer (basically a large protractor) and is usually described in degrees. If you can, while standing, lift your leg out in front of you so that it is parallel to the floor while straight (you would be at 90 degrees of hip flexion). This ROM is dependent on both the ability of the hip flexors (the agonist) to create hip flexion as well as the hamstrings (the antagonist) to lengthen. However, even more factors are at play.

ROM can also be further described as active ROM or passive ROM. The example in the previous paragraph uses active ROM and is a variation of Extended Hand-to-Foot Pose (Utthita Hasta Padangusthasana; figure 1.7*a*). Imagine this same pose but starting in a lying-down position. The pose then becomes a variation of Reclining Hand-to-Foot Pose (Supta Padangusthasana; figure 1.7*b*). It is essentially the same pose but with a different relationship to gravity. In the reclining version, if you take hold of your foot and pull the leg toward your chest, the arms generate a force that deepens the stretch on the hamstrings and allows you to explore your passive ROM. Additionally, if the leg is at more than 90 degrees of hip flexion, then gravity will help to bring the leg closer to the chest, just like the Leaning Tower of Pisa. Adding a bind or changing your relationship to gravity will affect whether you are using active or passive ROM.

Our passive ROM will always be greater than our active ROM, not just in a hamstring stretch but throughout the body. Consider a seated twist (figure 1.8). If you rotate your spine without using your hands to push yourself deeper into the twist, you might find you can rotate your trunk about 30 degrees, which is within the bounds of normal ROM. This rotation would come from your core musculature with the abdominal obliques playing a major role. If you then force yourself deeper into the twist using

a
b

FIGURE 1.7 *(a)* Active ROM, mostly limited by the strength of the hip flexors, in Utthita Hasta Padangusthasana. *(b)* Passive ROM, mostly limited by the stretch tolerance of the hamstrings, in Supta Padangusthasana.

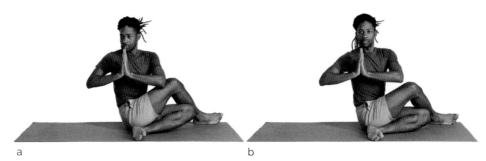

a   b

FIGURE 1.8   *(a)* Passive and *(b)* active versions of a seated twist.

your elbow against your knee to rotate or using a hand on the floor behind you, you would very likely rotate farther, perhaps to 45 degrees of thoracic rotation.

Many movement teachers, yoga teachers included, have begun to focus on active ROM over passive ROM. As mentioned previously, this might also be described as prioritizing mobility over flexibility. While there is crossover between the two, the development of active ROM requires a different approach to movement than the development of passive ROM. Consider Side Lunge (Skandasana). How low you can sit in Side Lunge would be a measure of your passive ROM in terms of hip flexion, knee flexion, and ankle dorsiflexion. How low you can transition into and out of your Side Lunge without using your hands on the floor would be a measure of your active ROM in those same joints. One way you can increase your active ROM in a pose is by having fewer points of support (e.g., try hovering your bottom hand above the floor in Triangle) or by transitioning in and out of a pose with control.

What Affects Flexibility

Have you ever stopped to wonder what controls your flexibility? In Reclining Hand-to-Foot Pose, what stops your leg from coming all the way up to your nose? Why can some people do the splits and others cannot? Also, what happens inside our bodies when we become more flexible? Many studies and reviews of studies have found that stretching can in fact increase flexibility (Behm, Blazevich et al. 2016; Freitas et al. 2018), but the mechanism behind why is not so clear.

The idea that flexibility comes down to muscle length is woven into our language and teachings. Most of us probably assume flexibility depends on muscle length, so by stretching a muscle, we lengthen it. Have you ever heard someone say they have short hamstrings? Physical therapy texts even describe techniques for measuring muscle length, though they are actually measuring joint angle and ROM (Weppler and Magnusson 2010). That distinction is an important one: Thinking ROM is due to muscle length is a false assumption, research shows. Consider how ROM is determined in living subjects. It is not by cutting them open, surgically removing their muscles, and measuring the length when stretched. In living subjects, the only way we can measure their flexibility is from their own report of sensation. We must stop stretching a subject if they say they cannot tolerate any more stretch. Thus, every measure of ROM depends on subject sensation and tolerance (Weppler and Magnusson 2010).

But if flexibility were simply down to muscle length, how could we explain the phenomenon that people become considerably more flexible under general anesthesia—

to the point that medics are trained in moving anesthetized patients so as to avoid dislocating their limbs (Baars et al. 2009)? A study by Krabak and colleagues (2001) examined the passive ROM in patients before, during, and after anesthesia, finding a significant increase during anesthesia. Knowing that anesthesia puts the nervous system in a suspended state, this suggests the CNS, not muscle length, is an important contributor to flexibility.

The CNS processes large amounts of data from our senses such as our eyes, our nose, and our ears, but also from our proprioceptors, which sense joint position as well as compression and tension of tissues. One such proprioceptor is the muscle spindle, a multitude of which are interwoven around muscle fibers, sensing tension on the muscle tissue and its surrounding fascia. As the muscle spindle senses a muscle being stretched to its end range, it sends a message to the spinal cord. The CNS reacts with a motor neuron telling the muscle to contract to avoid being overstretched and possibly injured. This is known as the *stretch reflex*, or *myotatic* (which means muscle tension) *reflex*.

The point of the stretch reflex is to keep us safe from injury. Without a stretch reflex, we could end up dislocating a shoulder by simply bumping into a pole. Recall how patients under general anesthesia are at much greater risk of dislocation. This is probably because their stretch reflex is not functioning fully or at all.

Normal movements that we make throughout the day are considered safe by the CNS because it is accustomed to these body positions. But when we put our body through a new movement or ROM, the CNS will put on the brakes by contracting the muscle to the range that it is accustomed. Some yoga textbooks cite the stretch reflex as responsible for limiting muscle extensibility, but experimental evidence has shown stretch reflexes to activate during very rapid and short stretches of muscles in a midrange position, producing a muscle contraction of short duration (Chalmers 2004). Additionally, stretched muscles do not exhibit a significant neural drive, meaning that the CNS is not telling them to actively contract (Magnusson et al. 1996). It seems that increases in muscle extensibility from stretching are more likely to be the result of alterations of sensation only, not muscle length, but much remains unknown about this interesting topic (Weppler and Magnusson 2010).

The Physiology of Adaptation

Chances are you have heard someone ascribe an injury to wear and tear or just old age. Undeniably, we are all aging, and aging brings various changes to the body. While a person's age should be considered in devising an appropriate movement practice (which, by the way, is not to say that an older adult should never do vigorous or strength-building exercise!), seeing the body as something that must be protected from wear and tear can have profound implications.

A physiological principle that should be taught in every yoga teacher training (and arguably to every human in the world) is that our bodies have an inherent and profound ability to adapt. While comparing a human body to a car can be a useful analogy at times (both burn fuel to create movement, etc.), there is one major difference between the two: a car cannot adapt to the demands placed on it, and the best way to preserve a car is to keep it in a protected garage and drive it very little. The opposite is the case for preserving the overall health of a human.

Wolff's law, as explored earlier in this chapter, states that bone adapts to mechanical loading. But this adaptive ability extends to many other tissues as well. Davis's law, the corollary to Wolff's law, describes how soft tissues—including muscle, tendons, ligaments, fat, fibrous tissue, skin, lymph and blood vessels, fasciae, and synovial membranes—model according to the demands placed upon them. All of these tissues adapt, meaning they can become stronger, more flexible, and more resilient. But they can also, if not loaded adequately, become weaker, less flexible, and less resilient.

Other tissues and systems can also adapt. The nervous system can adapt to learn new tasks or become better and more efficient at performing a task. It can even restructure its neurons to improve the way it delivers motor messages to muscles. The cardiovascular system can adapt to exercise to become more efficient at absorbing and transporting oxygen as well as removing carbon dioxide. Tissues can become more efficient at receiving oxygen. On a micro level, cells can become more efficient at producing energy from fat stores. These are only some of the many ways in which the body can adapt. Over the long term, all of these adaptations can improve health by reducing the risk factors for cancer, cardiovascular disease, type 2 diabetes, and many other diseases.

The body adapts very well to movement and loading. Though cartilage might not be able to remodel extensively due to its lack of blood flow within the joint capsule, it still benefits from movement and loading. Without your realizing it, your body is constantly adapting to the everyday needs that your life dictates. If you had to walk up three flights of stairs to reach your apartment, your body would quickly adapt to that demand. If, however, you were on bed rest for a few weeks, perhaps because of an illness, you might then find your first climb up the stairs afterward more challenging. You might feel your legs fatigue more quickly; you might be out of breath more than usual. This is the effect of detraining—even if your only training is normal everyday activities.

Progressive overload describes a method of training whereby the stress placed upon the body is gradually increased to stimulate muscle growth and strength gain. Muscles get stronger, of course, which is known as *hypertrophy*, but connective tissue and bones also respond favorably. Knowing that reduced muscle mass, known as *sarcopenia*, is associated with aging and lack of physical activity, we can all benefit from keeping strong and active even into old age.

An obvious example of adaptation with progressive overload is weightlifting, where a gradual increase in volume (number of repetitions) and intensity (percentage of maximum capacity) are the defining aspects. But does the same apply to yoga?

A load is a load, and yoga provides many opportunities to load the body in a variety of ways. You have probably experienced the effect of adaptation in the body. You might recall struggling with a certain yoga pose, then that pose becoming more comfortable as you practiced it. Or perhaps you felt your first strong yoga class was difficult to keep up with, but, over time, a similar level of class became doable.

Though the practice of yoga does not tend to incorporate any resistance beyond the practitioner's own body weight, yoga still offers an adequate stimulus for adaptation. Newcomers to yoga are often surprised at how challenging the practice can be and that yoga can even elicit delayed-onset muscle soreness, a sign of strength-building stimulus. Because a yoga practice is ever changing (unless a strict sequence is religiously followed), there are always new ways to effect change in the body.

If adaptations were to plateau, however, there are still many ways to progress a yoga practice. One could, of course, increase the frequency of one's practice (remembering though that adequate rest is important in the adaptation process). One could increase the intensity of the practice by simply holding poses longer, using repetition of poses, following a stronger style, or focusing on poses that engage many muscle groups, such as standing poses like Warrior II (Virabhadrasana II). Though yoga provides a variety of ways to build strength, flexibility, and resiliency, yoga practitioners still might benefit from adding strength training sessions outside of their yoga practice, where they lift loads beyond their own body weight by using dumbbells, barbells, or resistance bands.

While certain long-term health conditions such as multiple sclerosis or muscular dystrophy affect normal adaptation, we retain the ability to adapt throughout life. (And even with multiple sclerosis and muscular dystrophy, an individualized exercise program is still beneficial and recommended [Giesser 2015; Muscular Dystrophy UK 2015].)*

The idea that exercise (including yoga) might damage tissues through wear and tear can be harmful because it can discourage people from moving and loading their tissues, some of the beneficial things we can do for our body. Rather than talking about wear and tear, we consider movement and loading as wear and repair. While frequency and intensity of exercise do matter and ongoing health conditions must be considered, remember that as long as you are alive, you can retain the power to adapt, become stronger, and become more resilient.

Reciprocal Inhibition

A concept frequently cited in yoga textbooks and yoga classes is the principle of reciprocal inhibition, which describes the neurologic process of muscles on one side of a joint relaxing to accommodate contraction on the other side of that joint. Joints are controlled by two opposing sets of muscles that must work in synchrony for smooth movement. When a muscle is stretched and the stretch reflex is activated, an inhibitory interneuron in the spinal cord sends a message to the opposite muscle group to relax. This happens constantly to create smooth joint movements. During walking or running, as the quadriceps and hip flexors are activated to swing your leg forward, the hamstrings are inhibited until the leg nears the end of its forward swing when the hamstrings are activated to slow down and stop the swing.

Somewhere along the line, an assumption was made that reciprocal inhibition, when applied to stretching, should help to increase ROM. Many still claim today that by contracting the quadriceps during a forward fold, the hamstring will relax and allow for a deeper stretch, thereby increasing ROM. However, this idea has no scientific support (Sharman, Cresswell, and Riek 2006). In fact, the opposite has been shown to occur: One study found that contracting the quadriceps prior to a stretch to utilize reciprocal inhibition was less effective than static stretching alone for increasing hamstring extensibility over four weeks (Davis et al. 2005).

It seems that reciprocal inhibition is an instantaneous mechanism that allows for smooth coordinated movements between opposing muscle groups but is not important as a tissue-lengthening mechanism.

---

* While nearly everyone can benefit from some form of exercise, a few diseases, such as glycogen storage disease IV and V, necessitate careful dosage of exercise under medical supervision. You should always consult a physician before undertaking any exercise program.

TRY IT YOURSELF: A Qualitative Stretching Experiment

Instead of taking someone else's advice about the best way to stretch, why not decide for yourself? The following experiment uses Reclining Hand-to-Foot Pose (Supta Padangusthasana) because it allows the subject (you) to lie down and relax, thus reducing other confounding variables. You can try this with other asanas as well.

Lie on the floor and take a minute to relax, then try the following:

- *Active Stretching and Active ROM.* We often stretch our right side first, but why not start with the left? Lift the left leg high to test your active ROM. Keep the leg very straight. If your leg can be vertical (90 degrees of flexion) or closer to the chest, gravity will be helping you to deepen the stretch. If you cannot bring your leg to 90 degrees of flexion or more, you will be fighting gravity. Just notice where you are on the spectrum. This is active ROM.

- *Passive Stretching and Passive ROM.* Place a yoga strap around your foot and try to let the quadriceps (the antagonists to the hamstrings—the target of the stretch) relax as much as possible. Consciously relax the quadriceps, though they might need to be slightly engaged to keep the knee extended. Stay in this position for a minute or so. Notice how the hamstrings feel and how you feel in general.

- *Reciprocal Inhibition Stretching.* Try the method espoused by the reciprocal-inhibition-deepens-stretching camp. Engage the quadriceps strongly as you stretch the hamstrings. Stay in this position for a minute or so and continue to engage the quads the entire time. Notice how the hamstrings feel and how you feel in general.

- *Proprioceptive Neuromuscular Facilitation Stretching.* Try a proprioceptive neuromuscular facilitation technique by alternating between passively holding and actively contracting against a stretch. Still in Reclining Hand-to-Foot Pose, passively hold the stretch for a few seconds, then contract the hamstrings without moving (an isometric contraction) by pushing gently against the stretch without actually moving for 5 to 10 seconds. Then relax into the passive stretch again. Repeat this pattern twice more.

Complete the same sequence on the other leg (the right leg if you followed our guidance) and note any differences between the two sides.

Finally, pause to reflect on how the different stretching techniques felt. Did the reciprocal inhibition stretching make any difference to the passive stretch? How did the proprioceptive neuromuscular facilitation technique feel, both physically and energetically? Did one make you feel calmer? Which type of stretching did you prefer?

There is not one right way to stretch, and different techniques can be used at different times. By exploring different techniques on your own, you will be better equipped to guide others in doing the same.

INJURIES AND CONDITIONS OF THE MUSCULOSKELETAL SYSTEM

Though some media have portrayed yoga as an injurious activity, it remains a low-load physical activity with a relatively low risk for injury. Wiese and colleagues (2019) looked at self-reported yoga-related injuries in a large cross-sectional survey of 2,620 yoga participants, finding that severe injury was quite infrequent (4 percent of the total sample), which is relatively small given injury rates of other physical leisure activities. Parker and colleagues (2011), for example, reported at least 27 percent of women training to run a marathon experienced a severe injury during training. Among cheerleaders, Shields and Smith (2009) reported that 19 percent of those surveyed had experienced a severe injury (such as concussion, dislocation, or fracture) from their sport. The findings of Wiese and colleagues (2019) were consistent with previous reports that found serious injury to be a rare occurrence in yoga (Cramer, Krucoff, and Dobos 2013; Penman et al. 2012). Nonetheless, as is the case with any physical activity, yoga is not and cannot be a completely harmless activity. Some issues of particular interest in yoga are osteoporosis, hypermobility, stiffness, and low back pain.

The Goldilocks Principle and Injury

While the power of adaptation can make our tissues less susceptible to injury, everyone still has the potential to become injured, of course. What causes injury in one person, however, might cause no harm to another.

Quite simply, injury occurs when a mechanical load exceeds the strength and tolerance strength of a tissue. Thus, the Goldilocks principle is in effect: Too little mechanical stimulation leads to the weakening of muscles, bones, and other tissues, while too much can lead to injury. But if the stimulus is enough to challenge a person progressively (just right), then adaptations are favorable.

Though injury through yoga is possible (as it is with any physical activity), it is important to remember that yoga is a low-load, low-risk activity that is considered at least as safe as regular exercise (Cramer et al. 2015), and most yoga injuries are mild and transient (Cramer, Ostermann, and Dobos 2018). Furthermore, yoga is performed at a slow pace, and participants are regularly encouraged to move at their own pace and modify as needed. So, while a conversation around yoga-incurred injuries is worthwhile, we should also remember that it is a safe practice and potentially very beneficial to our overall health and well-being. A much greater risk to our health is being sedentary, and being too sedentary is known to be connected to cancer, cardiovascular disease, diabetes, and early death (Biswas et al. 2015).

For Goldilocks, finding the porridge of the right temperature was easy. Determining the right level of challenge is not so obvious. It is possible that on a particular day, the loading of a tissue might be too much, and injury occurs. But we can learn from every injury so as to hopefully avoid doing the same in the future.

Osteopenia and Osteoporosis

Osteoporosis (coming from Greek for porous bone) is a skeletal disorder characterized by low bone mass density making bones more susceptible to fracture or breaking. Osteopenia (Greek for bone poverty) is a condition of reduced bone density but not to the level of osteoporosis. Osteoporosis can be considered a silent disease, because people with it often have no symptoms and may not even know they have the disease until they break a bone. As the global population ages, osteoporosis is emerging as a significant public health problem with women over 50 being more likely to die from complications arising from osteoporotic fractures than from breast cancer (U.S. Department of Health and Human Services 2004). Although osteoporosis is most common in postmenopausal women, it is not limited to that population. In fact, men account for 30 percent of osteoporotic hip fractures worldwide (Cooper, Campion, and Melton 1992) and mortality after such fractures is greater in men than in women (Diamond et al. 1997).

Dual-energy X-ray absorptiometry is considered the gold standard for measuring bone mineral density (BMD). Osteoporosis is diagnosed when BMD is less than or equal to 2.5 standard deviations below that of a healthy 30- to 40-year-old adult female reference population. Described in terms of a T-score, the World Health Organization has established the following diagnostic guidelines (Genant et al. 1999):

Normal. −1.0 or higher

Osteopenia. −1.0 to −2.5

Osteoporosis. −2.5 or lower

Severe Osteoporosis. −2.5 and already incurred a fracture due to bone fragility

Many concerns about osteoporosis and safety swirl about the yoga community. Teachers, for example, often caution students with osteoporosis to avoid all twists and forward folds. But is this the best advice for yoga practitioners with osteoporosis?

Spinal Flexion and Extension

There is some evidence that spinal flexion might be contraindicated for people with osteoporosis. Before the mid-1980s, spinal flexion exercises (e.g., forward bending with a rounded spine or abdominal crunches from a supine position) were often recommended to alleviate back pain related to vertebral fractures (Sinaki 2007). However, in 1984, Sinaki and Mikkelsen showed an association between spinal flexion exercises and an increased incidence of vertebral fractures in osteoporotic women (Sinaki and Mikkelsen 1984). Sinaki (2013) also documented case studies of three women (ages 61, 70, and 81) with low BMD who experienced pain and new fractures, which two of the women attributed to Plow Pose (Halasana) and the last attributed to this pose plus Shoulder Bridge (Setu Bandhasana). These findings together suggest that a yoga practice or exercise regimen characterized exclusively by spinal flexion exercises might increase the risk of spinal fractures for women with osteoporosis.

Conversely, strengthening the spinal extensors has been seen to benefit people with osteoporosis. Excessive thoracic kyphosis (i.e., hunchback) can be an independent risk factor for fractures and several studies provide evidence that strengthening the spinal

Can Yoga Improve Bone Health?

It is well established that physical activity strengthens bones. As you walk, for example, you exert up to 1.5 times your own body weight onto the ground, and the ground reacts with an equal amount of force, which is then absorbed by the body. When running on a treadmill, two to three times your body weight is absorbed by each foot with every step (Kluitenberg et al. 2012). However, does the practice of yoga asanas provide adequate stimulus for strengthening of bone?

The mechanical stimulus required to remodel bone is not an all-or-nothing matter. Every moment spent in an environment with gravity affects bone health. Even seated, as you probably are right now, your skeleton is being compressed by gravity and responding as it should, putting down new layers of bone to meet the demands of living on Earth.

This means that if someone went from complete bed rest to standing, their leg and hip bones would become stronger. If that person then began to walk, those same bones would eventually become stronger still. And if that individual eventually began running, the lower limb bones would become even stronger. Of course, this progression from complete bed rest to running would have to be gradual and done intelligently with adequate time to rest and remodel.

Beyond weight-bearing and ground-reaction forces, the pull of muscle on bone is sufficient to create adaptation, so the contraction of your shoulder muscles to hold your arms in Warrior II may be sufficient to generate remodeling. Stretching, which involves the pulling of muscles on the periosteum of bone, may even be sufficient to elicit adaptation, at least in people who are not conditioned to stretching.

A beneficial aspect of yoga asanas is bearing weight in a variety of different ways. Of course, we bear weight through our feet in standing poses such as Tree (Vrksasana) and Triangle (Trikonasana). But we also bear weight through our hands in Downward-Facing Dog (Adho Mukha Svanasana) and Low Plank (Chaturanga). Even bearing weight through the head in Headstand (Sirsasana) can increase bone strength of the skull and cervical vertebrae. (But there are many factors to consider in determining whether Headstand is appropriate for a practitioner, and such a discussion is beyond the scope of this book.)

In a study published in 2016, Lu and a team of researchers including well-known author and yoga advocate Loren Fishman looked at whether yoga asanas can positively affect BMD in people with osteoporosis (Lu et al. 2016). The researchers recruited 1,000 people from all over the world and asked them to follow a digital video disc that provided a 12-minute yoga practice with 12 poses. Eight years later, over 240 of the recipients complied and sent in previous dual-energy X-ray absorptiometry scans.

The results were surprising. Over 80 percent of the people in the study reversed their bone loss and began to gain bone. Importantly, no fractures or serious injuries of any kind were seen or reported in over 100,000 hours of people performing this yoga sequence daily. Over 80 percent of the subjects had osteoporosis or osteopenia at the beginning of the study, and fewer had these conditions by the end. As with any experiment, this study had its limitations. Most importantly, the subjects were at home, not in a controlled environment over a long period of time, thereby increasing the chances that confounding variables could have also affected the results. However, this study provides novel information that reversal of bone loss in the spines of people with osteoporosis is possible and that these 12 asanas, practiced daily, provide adequate stimulus to generate bone strengthening.

extensors is associated with decreased thoracic kyphosis (Itoi and Sinaki 1994; Sinaki et al. 2002) as well as improved quality of life (Hongo et al. 2007). Most importantly, some evidence has shown that strengthening the back extensors may provide long-term protection against vertebral fractures, independent of bone mineral density (Sinaki et al. 2002).

Spinal Rotation and Lateral Flexion

Though the previously mentioned studies provide some helpful information on exercise for people with osteoporosis, there have been very few other studies exploring the most beneficial exercises for people with compromised bone density (Pratelli, Cinotti, and Pasquetti 2010). Very little empirical evidence exists about the effects of spinal rotation (twisting) and lateral flexion (side bending) on people with osteoporosis (Smith and Boser 2013). From the perspective of scientific literature, there is no basis to informing people with osteoporosis to avoid spinal rotation or lateral flexion.

To the contrary, varied movement and dynamic loading are essential to the health of the spine and its intervertebral discs. Movement and loading help to regulate bone density and to move nutrients and waste products into and out of the cells of the spine and the discs (Chan et al. 2011). Discs are avascular, meaning that they have no blood flow, and so instead rely on diffusion to move needed nutrients throughout their cells. Engaging in physical activity and frequently changing body positions promote the flow of fluids to and from discs (Chan et al. 2011). Furthermore, bone is anisotropic, which means that its physical strength varies along different axes. To make bone its strongest, we should dynamically load it in different ways, along different axes. If we were to stop all rotational movement in the spine, it would not have the stimulus to grow stronger along that axis, the transverse axis. This means that when it comes time to rotate the spine in real life, the spine will be less conditioned to support that rotation and might be more likely to incur injury. As with the rest of our body, it seems that the spine and the discs benefit most from dynamic, moderate, weight-bearing exercise (Smith and Boser 2013).

Hypermobility

Hypermobility has begun to garner attention in the yoga community—and for good reason. With social media full of images of yoga influencers twisting themselves into highly contorted positions, yoga can appear to be about ever-increasing and near-superhuman flexibility. With its focus on end-range mobility, asanas tend to attract exceptionally flexible individuals. While most yoga practitioners would surely agree that yoga is for everyone, being good at yoga can sometimes be construed with getting oneself into poses that resemble human pretzels. It is worthwhile to ask: Do yoga asanas promote an unhealthful level of flexibility and should hypermobile individuals practice yoga?

Joint hypermobility, commonly known as *double-jointedness* (though that term is not anatomically or biomechanically accurate), describes joints that can move beyond what is considered normal ranges of motion. While the definition of normal range of motion varies from one textbook to another, a few indicators of hypermobility are the ability to touch your thumb backward to your wrist or the ability to put your foot behind your head. The latter is a position that, on the one hand, is considered abnormal by the medical community and is, on the other hand, known as *Eka Pada Sirsasana* and found in the intermediate series of the Ashtanga Vinyasa system.

While it is possible to be hypermobile in just one joint, hypermobility is usually seen in multiple joints or even all joints as a hypermobile individual has highly compliant (meaning stretchy) connective tissue. Joint hypermobility is relatively common, occurring in about 10 to 25 percent of the population, with no problems presenting in most cases (Garcia-Campayo, Asso, and Alda 2011). If no symptoms are present, there appears to be no harm in being hypermobile.

A minority of hypermobile people, however, can experience pain and other issues. *Hypermobility spectrum disorder* (HSD) describes the pain and other symptoms that can come with being overly mobile. The term was coined in 2017 to replace *joint hypermobility syndrome*, which lacked differentiation from other syndromes (Tinkle et al. 2017). Symptoms of HSD can include pain in affected areas and the inability to walk properly or for long distances. Some people with HSD have hypersensitive nerves and a weak immune system. It can also cause severe fatigue and, in some cases, is associated with depressive episodes or anxiety disorder. It is similar to other genetic connective tissue disorders such as Ehlers-Danlos syndrome. In fact, some experts recommend the two should be recognized as the same condition until further research is conducted, because no genetic test can identify or separate the conditions and the diagnostic criteria and recommended treatments for them are similar.

A common diagnostic tool for measuring hypermobility is the Beighton test. It is a nine-point scale that primarily looks at a person's passive range of motion in a few joints: both thumbs, both little fingers, both elbows, both knees, and the trunk (figure 1.9). Different criteria have been used to measure the results of the tests, ranging from more than three hypermobile joints to more than six hypermobile joints of the nine assessed, but the most frequent choice of cutoff is more than four hypermobile joints (Clinch et al. 2011).

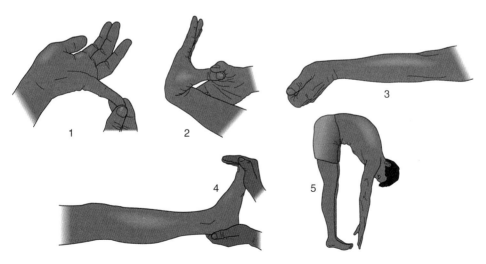

FIGURE 1.9 The five maneuvers of the Beighton test.

Having been in use for more than 30 years, it is a quick and easy way of measuring hypermobility. However, this test looks at only a small number of joints, so hypermobile joints outside this selected group are overlooked. Also, the test does not include other systems that can be affected by hypermobility, including the digestive system. It also provides only a yes-or-no result and does not indicate degree of hypermobility.

Hypermobile people might be more drawn to the practice of yoga than hypomobile (very stiff) people as they find they can access the advanced postures with relative ease. But should a hypermobile person, especially one who experiences pain as a result of their increased flexibility, really be trying to increase their flexibility through yoga?

While a full discussion around hypermobility and yoga is beyond the scope of this book, we believe yoga can be appropriate for hypermobile people, whether diagnosed with HSD or not. The intricacies of adapting yoga to hypermobile populations create a complex topic, and what works for one hypermobile person may not work for another. Furthermore, yoga teachers do not have the authority to diagnose HSD—or any condition for that matter. However, with those caveats, here are a few common-sense guidelines for hypermobile people practicing asana:

- Maintain a slight bend in joints, particularly weight-bearing ones. Hypermobile people are more likely to have joints that can hyperextend (one of the marks from the Beighton test is hyperextended elbows), and it is easy to disengage muscles around hyperextended joints, which usually translates as greater force through ligaments and joint capsules. In addition to maintaining a microbend in the elbows during High Plank, try microbending the knee in straight-legged poses such as Triangle (Trikonasana) and Pyramid Pose (Parsvottanasana). Also, press into the ball of the foot (metatarsals) while driving the shin forward slightly. This will require more effort but can contribute to building strength in the asana.

- Prioritize stability over flexibility. While passive stretching can feel great and can complement a balanced movement practice, hypermobile people should probably focus on increasing strength and stability over increasing range of motion. Knowing that hypermobile people usually have decreased muscle strength and muscle mass (Pacey et al. 2010), training for strength should be an important part of their practice. In any given pose, instead of focusing on going deeper into the stretch, consider how easily you could be knocked over if someone bumped into you and focus on creating more stability in your practice rather than flexibility. In making strength a priority, one might consider adding a session of resistance training with weights or resistance bands to their weekly yoga practice.

- Aim for 80 percent of your full range. Pulling back from your end range will make your muscles recruit to support the pose and may assist in building your body's inner awareness, or proprioception, something that tends to be compromised in hypermobile people.

- Seek ideal individual alignment rather than ideal universal alignment. Notice how the traditional cues for an asana (which can vary from one style to another) feel in your body. If the teacher is telling you to keep the heels in one line in Warrior II while squaring off your hips to the front of your mat, but you find that these cues do not work for your body (for example, they cause pain in the lower back), be confident in adapting the pose to your body's needs. Sometimes, a slightly wider foot stance or shift in the hips can make all the difference between a pose feeling uncomfortable or comfortably challenging.

- Practice humility. Just because you can get your foot behind your head does not mean you should. While there are no poses that are specifically contraindicated for hypermobility, hypermobile people should consider the following question before trying to get into some contortionist-looking advanced posture: Am I attempting this pose because I genuinely believe it would benefit my body, or am I attempting it merely out of ego?

Stiffness

K. Pattabhi Jois, creator of ashtanga yoga, is credited with saying, "Body not stiff; mind stiff" (Mana Yoga 2011). The popularity of ashtanga yoga has, for many reasons, decreased immensely, but there still exists the idea that mental blockages manifest as tightness in the body. And when that quote by Jois is juxtaposed with a yogi peacefully posed with one foot behind their head, it is easy to wonder what blockages are keeping you from doing the same.

Tightness or stiffness lead many people to yoga. Many massage therapists have uttered the words, "Wow, you're really tight," and in our lexicon, we might tell someone who seems agitated to loosen up. But what is this stiffness or tightness that seems to affect every one of us at least occasionally?

First, *stiffness* and *tightness*—and other iterations of these words—are subjective terms to describe how someone feels in their body. On the other hand, ROM is an objective measure of how much movement is possible at a joint. Flexibility then is more accurately described in degrees of ROM.

Consciously or unconsciously, most of us probably attribute the feeling of stiffness to reduced ROM. But this connection is not so clear. In one study, Stanton and colleagues (2017) used a back-probing device on people with and without low back pain to measure whether feelings of back stiffness related to objective spinal measures of stiffness. Over three experiments, they found that feeling stiff did not relate to objective spinal measures of ROM. Additionally, actual reduced spinal ROM did not differ between those who reported feeling stiff and those who did not. Those who did feel stiff, though, exhibited self-protective responses. That is, they significantly overestimated the amount of force applied to their spine, yet they were better at detecting changes in this force than those who did not report feeling stiff. The researchers also experimented with synchronizing sound to forces being applied to the spine, finding that what the subject heard affected their perception of how much force was being applied. This study provided a compelling argument against the prevailing view that feeling stiff is an accurate marker of actual back stiffness. Rather, feeling stiff is a multisensory perception consistent with ideas of protecting the body.

Even stranger is that people can feel stiff and arthritic in joints they no longer have. Haigh and colleagues (2003) observed that three patients with rheumatoid arthritis continued to feel stiffness in joints of amputated legs. Also, the arthritis symptoms in their phantom limbs responded to the use of anti-inflammatory drugs in the same way that their actual limbs did. In other words, drugs that have been sold as targeting joint pain seem to work even when no target exists.

Additionally, people with small ranges of motion can feel fine while very flexible people can feel very stiff and tight. In fact, hypermobility is associated with persistent pain even with no perceivable tissue damage, which suggests that the connection with flexibility and sensation is a complex one (Scheper et al. 2015). A reduced ROM is probably never in itself the cause of any sensation of stiffness. Feelings of stiffness can even occur during normal movements in normal ranges of motion. Most of us have surely experienced a feeling of stiffness, perhaps in the low back, after an extended period of sitting.

So, the sensation of stiffness is not correlated with actual reduced ROM except perhaps in the morning. Many of us experience reduced ROM in the morning and usually an accompanying sensation of stiffness. But what about feelings of stiffness that seem to persist despite the time of day?

While feelings of stiffness can arise regardless of the state of our tissues, persistent sensations of stiffness can indicate minor pathology. Stiffness might be a precursor to pain, a kind of mild pain, and it can be as multifaceted and complex as pain (see the Persistent Pain section in chapter 2 on the nervous system). Therefore, figuring out the cause of stiffness can be as difficult as figuring out a not-so-obvious cause of pain. However, there are a few possible reasons for stiffness.

Delayed-Onset Muscle Soreness

Delayed-onset muscle soreness (DOMS) is the soreness experienced after a bout of exercise—even a vigorous yoga session—and usually appears the day after exercising, leaving the muscle more sensitive to touch and to movement. It is generally accepted that the training stimulus creates microdamage in the body, leaving a sensation that some describe as pain or discomfort, and that the damage repairs over the next few days.

Physically, DOMS can lead to a temporary reduction in the amount of force a muscle can generate, a disturbed sense of joint position, decreased physical performance, and an increased risk of injury (Dupuy et al. 2018).

DOMS is greater after exercise to which the body is unaccustomed than to exercise that it is accustomed to. DOMS seems to be greater after eccentric than concentric contractions but does still occur after concentric-only exercise. DOMS does not appear immediately after exercise but gradually after, and it reaches a peak between 24 and 72 hours postexercise. DOMS tends to occur more in certain regions of the exercised muscles than in others—for example, more at the distal ends of the quadriceps than the proximal ends, though this can vary among individuals.

DOMS can even be experienced after a bout of stretching. Have you ever found your hamstrings sore after a yoga class? If so, the stimulus from stretching them, though it may have been relaxing for your mind, was enough to create soreness.

Surprisingly, even after decades of research, we do not understand the physiology behind DOMS. Despite being frequently cited in popular media, the buildup of lactic acid in the muscles is not a viable explanation for DOMS. Although the accumulation of metabolites (including lactic acid, which is immediately turned into lactate and hydrogen ions) is at least partly responsible for the burn experienced during high-intensity exercise, including holding a Chair Pose for what feels like much more than five breaths, the production of lactate does not differ significantly between unaccustomed and accustomed exercise, as shown by the lack of any difference in blood lactate between trained and untrained subjects (Gmada et al. 2005). Plus, lactate is shuttled out of muscles very quickly, whereas DOMS takes hours or days to appear after exercise.

Beyond this old lactic acid theory, other possible mechanisms have been proposed to explain DOMS, including damage to muscle fibers, fascia, or muscle spindles of the nervous system. It might even be from the inflammation or swelling of muscle fibers and fascia, or even from oxidative stress within muscle fibers. How nerve changes (through neurotrophic signaling) might fit in is not clear. All these theories are viable explanations to DOMS, but clearly there is a lot that we do not yet know (Beardsley 2020).

Muscle Knots

We have all probably experienced what is commonly called a *muscle knot*. Of course, a muscle cannot tie itself into a knot, so what can explain this sensation? We often seek massage for these knots, and applying pressure on them seems to provide relief.

A muscle knot is sometimes called a *trigger point*. In *Myofascial Pain and Dysfunction: The Trigger Point Manual* (1983), Janet Travell and David Simons define a trigger point as an irritable nodule, often palpable, in the taut bands of fascia surrounding skeletal muscles. They note that direct compression and muscular contraction on the point can incite a startled response (jump sign), local tenderness, local twitch response, and referred pain somewhere distant to the trigger point. As the title of their book suggests, they attribute these trigger points to microcramps in the muscle and issues with the fascia, at least in part.

The book achieved great commercial success, but it relied on many assertions and beliefs that were derived without testing or without adequate scientific basis. In 1992, Wolfe and colleagues performed a study on trigger points. They recruited a group of four myofascial pain experts, selected by Simons himself and including Simons, who examined four patients with myofascial pain. The examiners could take as much time as they needed to examine but were not allowed to interview the patients. The four patients were controlled against healthy subjects as well as subjects diagnosed with fibromyalgia (discussed later) without the examiners knowing who was who.

Even though these trigger point experts were the best—Simons cowrote the book on it—they could not find or agree on the trigger points. The lead author of the study, Dr. Fred Wolfe, later revealed in a blog post:

> It was a disaster. The examiners were distraught. After the results were in, they protested and wanted to change the protocol and purposes of the study (post hoc). It wasn't fair, they said . . . If we believed in trigger points and The Trigger Point Manual before, we were a lot less secure in our beliefs now. (Wolfe 2013)

Despite the lack of scientific basis behind it, the myofascial pain theory still dominates, even among professional circles, as an explanation of muscle knots.

Other scientists have tried to explain the phenomenon of muscle knots. Quintner and Cohen (1994) suggested that irritated or injured peripheral nerve trunks, rather than microdamage in muscle tissue, may be the cause of pain. Like the idea it is intended to replace, this hypothesis has advantages and problems; the main problem is that there is no obvious plausible mechanism for pervasive nerve irritation.

The reality is, to this day, we cannot clearly explain what a muscle knot is or the mechanism behind it. Yet few would dispute the existence of muscle knots because we have probably all had one at some point. Whatever the mechanism behind these pesky irritants, here are a few methods worth trying for relief from them:

- Resting
- Stretching
- Exercising
- Hot and cold therapy
- Self-massage or massage therapy
- Physical therapy, particularly if the problem lingers, creates discomfort, and does not respond to treatments at home

Fibromyalgia

Fibromyalgia is a long-term disorder characterized by widespread musculoskeletal pain or stiffness and extreme fatigue accompanied by sleep, cognitive, and mood issues. The exact cause of fibromyalgia is unclear, but it may be related to injury, emotional distress, or viruses that change the way the brain perceives pain. No diagnostic test for fibromyalgia exists, so health care providers diagnose it by examining the patient, evaluating symptoms, and ruling out other conditions.

Symptoms often begin after an event, such as physical trauma, surgery, infection, or significant psychological stress such as the breakdown of a relationship or the death of a loved one. In other cases, symptoms gradually accumulate over time with no single triggering event. Women are more likely to develop fibromyalgia than men and many people who have fibromyalgia also have tension headaches, temporomandibular joint disorders, irritable bowel syndrome, anxiety, and depression.

Given that fibromyalgia is unexplained, much controversy and many unfounded theories surround the condition. But a few treatment options exist with some good efficacy, including exercise, relaxation, and good sleep hygiene (NCCIH 2016).

Conclusion About Stiffness

Stiffness and *tightness* are vague terms that mean different things to different people. Many assumptions have been made about stiffness, and many products have been sold with a promise of relieving stiffness, though most are not founded in good science. Stiffness might be associated with reduced extensibility of a muscle and connective tissue, but it could also be associated with a dull ache in the region, muscle knots, or any other sort of mild pain. The potential causes of stiffness are many, and some of the time (maybe much of the time), the exact cause cannot be named.

We should also remember that the human body is a tensile structure, held together and able to move well because of a balance of tension and compression in a concept known as *tensegrity*. Tension helps us resist gravity and move in the many ways we do. While the words *tension* and *tightness* usually carry a negative meaning, they play an important role in making this body of ours functional.

Low Back Pain

Being the number one cause of disability in the world, low back pain (LBP) is indeed a major problem of our contemporary world (Hartvigsen et al. 2018). At some point in their life, an estimated 60 to 80 percent of adults will suffer from some form of LBP, and in a surprising 85 percent of cases, LBP is not attributable to a specific pathology (Airaksinen et al. 2006). These remarkable numbers show that most of us will experience LBP at some point, and there will be no identifiable cause for it.

From these numbers, one might surmise that LBP is just a part of being human. There are, however, a few things we can do to prevent the onset of LBP and to rehabilitate from it. Many people turn to yoga to alleviate back pain, and yoga is a recommended activity of many health care providers. It is a certainty that all yoga teachers will have students in class who currently have LBP or have had it in the past. The most important thing we can do to improve our back pain or help others understand theirs begins with language and mindset.

Many unhelpful beliefs about back pain are common among the general population, the media, yoga teachers, and even medical professionals. And beliefs are powerful. Unhelpful beliefs about LBP are associated with greater levels of pain, disability, work absenteeism, medication use, and health care seeking (Main, Foster, and Buchbinder 2010). Unhelpful beliefs can trigger what is known as the *nocebo effect*.

The Nocebo Effect and Fear-Based Language

Most people are familiar with the placebo effect, which describes a positive response to an inert stimulus. A classic example of this phenomenon, which has been observed in many studies, is an individual being suddenly freed from persistent pain after taking a so-called pain-relieving drug that is nothing more than a pill of starch. The word *placebo* is a first-person conjugation of the Latin *plācāre* and means "I will appease; I will placate." The placebo effect demonstrates the tremendous power of expectation. But what if an expectation is one of harm?

A 2018 review found that 49 percent of patients taking placebos in clinical trials experienced adverse events, such as headache or nausea (Howick et al. 2018). In other words, almost half of all people receiving an inert substance experienced negative effects. This appears to occur both when the subject is warned of possible side effects (as might happen in a double-blind trial where neither the subject nor the researcher knows who is having the trialed drug) and also when the patient is not warned of any possible side effects, whether the drug is real or not (Howick et al. 2018). It seems the power of expectation can go either way: favorably or unfavorably.

Coined in 1961, the term *nocebo effect* describes the phenomenon of experiencing an unfavorable effect from an inert stimulus. Coming from the Latin *nocere* meaning to hurt, harm, or damage, *nocebo* means "I will harm; I will damage." The nocebo effect occurs not just with the ingesting of an inert pill; it can also occur with a simple verbal suggestion (Benedetti et al. 2007). If people are told that something will hurt, they are more likely to experience pain. As for the mechanism behind this, negative verbal suggestions can induce anticipatory anxiety, which triggers the activation of cholecystokinin, a hormone that facilitates pain transmission (Benedetti et al. 2007).

Endogenous opioids and dopamine also play a role in pain perceptions and therefore in the placebo and nocebo responses (Benedetti et al. 2007). (Pain is a complex topic and further information can be found in chapter 2 on the nervous system.)

Certain factors might affect someone's likelihood to experience the nocebo effect. Individuals with anxiety and depression and those with a tendency toward somatization have been found to be more likely to exhibit the nocebo response (Wells and Kaptchuk 2012). According to Drici and colleagues (1995), type A subjects (aggressive, competitive, and driven toward achievement) described subjective side effects of an inert substance more than other personality types. Pessimism may also predispose people to negative expectations and to the nocebo response (Data-Franco and Berk 2013). However, under the right circumstances, we can probably all find ourselves susceptible to the nocebo effect.

The nocebo effect can be directly linked to yoga as well as our own self-perceptions. Consider whether you have ever heard any of the following cues in a yoga class:

- Your knee must be positioned above your heel in a lunge to protect your knee.
- If you don't engage your core, you'll injure your back.
- Engaging your glutes in Shoulder Bridge (Setu Bandhasana) compresses the lower back.
- The sacroiliac is a very vulnerable joint.
- The shoulder is susceptible to injury.
- There is only one right way of performing a pose.

While these cues are surely well-intentioned, they can elicit the nocebo effect. Consider the cue to "Engage your core to protect your back." What is implicated to happen if you do not engage your core? This cue, while well intended, suggests that our backs are fragile, and injury is more likely if we do not engage our cores. Engaging one's abdominal musculature, particularly during physically demanding poses like Backbends, is probably a good idea and a good cue to offer. However, language matters, and telling someone that they might injure themselves can plant a seed that can have unintended negative consequences.

If the study of physiology has taught us anything in the last century, it is that our bodies adapt. Our bodies are not just robust; they are, as Nassim Nicholas Taleb describes it, antifragile (Taleb 2012). If something is robust, it is strong, but once it is pushed beyond a certain point, it is permanently broken. Though any tissue of the body can of course be subject to injury or disease, our bodies respond positively to change. Do you want to share a message that our bodies are delicate and prone to wear and tear, or do you want to share an empowering message that throughout our lives, our bodies maintain the ability to adapt in positive ways? The latter is not just more empowering; it is also more accurate.

Terms to Use Instead of Fear-Based Language

Perhaps when teachers mean to protect, they simply mean to stabilize. Think of using language that is empowering. Rather than saying, "Stop if anything hurts," consider saying, "Just do what you can. Hopefully some movement will help." You would be correct in saying that movement and exercise are some of the best things we can do for our body, and science supports that claim.

We should also be careful about not being too prescriptive with our language. Two people can have the same injury but different aggravating triggers. With a herniated disc, for example, two people can have the herniation in the same place, yet one might feel discomfort during a forward fold (spinal flexion) while the other during a twist. Knowing that mechanical stimulation is necessary to remodeling and adaptation, let's strive to keep people moving. Remember that yoga is a low-load activity modifiable to a large variety of populations.

More than 75 percent of people have not heard of or do not believe in the nocebo effect, but learning about this phenomenon can help to reduce its negative outcomes (Planès, Villier, and Mallaet 2016). Therefore, awareness and recognition of the nocebo effect is a good first step.

Empowered to Move

While unhelpful beliefs can worsen LBP, the opposite is also true: a positive mindset around LBP is associated with lower levels of pain, disability, and health care seeking (Beales et al. 2015). Led by Peter O'Sullivan in Australia, a team of highly credentialed experts in pain and rehabilitation collaborated to write a paper and create a handout intended for public use on what we know about back pain from decades of research and clinical practice with the aim of empowering LBP sufferers (O'Sullivan et al. 2020). They wrote that, once red flags and serious pathology are excluded, evidence supports the following:

- LBP is not a serious, life-threatening medical condition.
- Most episodes of LBP improve, and LBP does not get worse as we age.
- A negative mindset, fear-avoidance behavior, negative recovery expectations, and poor pain coping behaviors are more strongly associated with persistent pain than is tissue damage.
- Scans do not determine prognosis of the current episode of LBP nor the likelihood of future LBP disability, and they do not improve LBP clinical outcomes.
- Graduated exercise and movement in all directions is safe and healthy for the spine.
- Spine posture during sitting, standing, and lifting does not predict LBP or its persistence.
- A weak core does not cause LBP, and some people with LBP tend to overtense their core muscles. While it is good to keep the trunk muscles strong, it is also helpful to relax them when they aren't needed.
- Spine movement and loading are safe and build structural resilience when done gradually.
- Pain flare-ups are more related to changes in activity, stress, and mood rather than structural damage.
- Effective care for LBP is relatively cheap and safe. This includes patient-centered education that fosters a positive mindset and coaching people to optimize their physical and mental health by engaging in physical activity and exercise and social activities, developing healthy sleep habits, achieving and maintaining body weight, and remaining employed.

While back pain can be frustrating and at times disabling, it is helpful to remember that 90 percent of LBP cases are self-limiting and resolve within six weeks (Waddell 1987). Many factors affect back pain, including our own perceptions of the pain itself. Knowing that our language can affect pain through the nocebo response, we are well advised to be aware of how we use our words. Finally, as with everything else in the body, one of the best treatments we can pursue for back pain is just what our yoga practice provides: movement.

CONCLUSION

The musculoskeletal system requires stress (mechanical loading) to make it strong and resilient. From preventing pathologies to keeping bones healthy, movement is one of the best things we can do right now for our bodies. While any tissue has the potential for injury, loading the body is all about finding the right balance between what is too easy and what is too challenging.

Chapter 2

NERVOUS SYSTEM

Yoga is often described as a mind–body practice. But what does this mean, and how does yoga affect the mind? The mind and the brain are not the same. The mind is generally associated with feelings, sentience, and consciousness of self, while the brain is a dissectible biological tissue. But learning about the brain, which is part of a larger system called the *nervous system*, can teach us about the mind and thus about our nature.

The nervous system is the most complex and highly organized system in the human body. It is our major controlling, regulatory, and communicating system and is the center of all mental activity including thought, learning, memory, and behavior. The nervous system keeps us in touch with our environment, both external and internal. Together with the endocrine system, the nervous system is responsible for homeostasis, which is the ability to maintain a relatively stable internal state that persists despite changes in the world outside. This system is responsible for controlling all voluntary movement (so, every physical action performed during a yoga practice) and, last but certainly not least, keeping us alive by constantly being on the lookout for anything that is perceived to be threatening.

The nervous system can be divided into central and peripheral. The central nervous system (CNS) consists of the brain, the spinal cord, and the retinas of the eyes. The peripheral nervous system (PNS) consists of all the components of the nervous system that lie outside the brain, spinal cord, and retinas.

Highlights of this chapter include how yoga can help us to manage modern-day stress, the latest research looking at yoga for mental health, and up-to-date information on persistent pain.

CELLS OF THE NERVOUS SYSTEM

The two major cell types that make up the nervous system are neurons and glial cells. Neurons generate and conduct electrochemical impulses, while glial cells provide the neurons with mechanical and metabolic support.

A Personal Note From Andrew

When I graduated from medical school in England in 2005, I was really concerned about how I was going to cope with the stress and demands of working as a junior doctor (the equivalent to an intern in the United States). A good friend highly recommended meditation, which at the time was something that I had heard about but did not completely understand. I booked a private session with an experienced local teacher, and at the very first practice I had an extremely profound, almost out-of-body, experience. Meditation quickly became one of my most treasured coping mechanisms for managing stress, and I practiced twice daily without fail for many years. My career as a doctor did not last long as I set my sights on other paths, but without the daily anchor of meditation, it would have been an even shorter chapter for me. As I move through my life, meditation remains a significant part of my toolbox, allowing me to be a more present husband, friend, son, brother, uncle, and teacher.

The main nucleus-bearing part of the neuron (figure 2.1) is called the *cell body*. The cell body is similar in structure to other cells in our body, but it does not have the same ability to regenerate after injury, posing significant issues for the treatment of injury and disease of the nervous system. Many of these cell bodies have projections called *dendrites*, which carry information to the cell body. The axon is the long structure along which the nerve impulse passes from the cell body. Most axons are insulated by a fatty substance called *myelin*. There are periodic interruptions where short portions of the axon are left uncovered by myelin; these are called the *nodes of Ranvier*. The junction between one axon terminal and the dendrite of another neuron is called a *synapse*. Synapses are often referred to as *points of connection* between neurons, but they are actually short gaps between the neurons in the same way that a joint is a junction and small gap between two bones.

Inside the resting axon is a low concentration of sodium ions and a high concentration of potassium ions compared to the surrounding tissue fluid. This results in the inside of the axon being more negatively charged than the outside. When stimuli are detected by the neuron, sodium ions enter the axon, and potassium ions leave the axon via channels at the nodes of Ranvier. This makes the inside more positively charged and results in an electrical charge (nerve impulse or action potential) traveling along the membrane of a neuron. The nodes of Ranvier are essential in the speed and timing of delivery of impulses from one neuron to another. At the synapse the transmission of a nerve impulse is passed on to another neuron. After the nerve impulse has passed, the resting state of the axon is restored. This whole process takes milliseconds with nerve impulses traveling up to 120 meters per second in humans, which is 268 miles per hour.

Neurotransmitters are endogenous chemicals, meaning they are produced within the body, and they allow neurons to communicate with each other across synapses. Different neurotransmitters are used by the body for different functions, including dopamine, serotonin, and histamine. Dopamine plays an essential role in several brain functions, including learning, motor control, reward, emotion, and executive functions (Ko and Strafella 2012). Serotonin is a neurotransmitter that modulates multiple

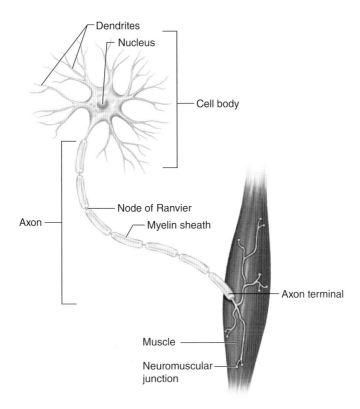

FIGURE 2.1 Structure of a neuron.

neuropsychological processes and also has implications that affect gastrointestinal processes such as bowel motility (Berger, Gray, and Roth 2009). Histamine mediates homeostatic functions in the body, promotes wakefulness, modulates feeding behavior, and controls motivational behavior (Passani, Panula, and Lin 2014).

The three types of neurons are sensory neurons, motor neurons, and interneurons. Sensory neurons convert external stimuli from the environment into corresponding internal stimuli. They are activated by sensory input such as light, sound, smell, taste, heat, and physical contact, and they convey this information to the spinal cord or brain. Motor neurons are involved in both voluntary and involuntary movements through the innervation of muscles and glands. The two types of motor neurons are upper motor neurons and lower motor neurons. Upper motor neurons originate in the brain, integrate all the signals received by the brain, and translate these into a single signal that either initiates or inhibits voluntary movement. They connect to lower motor neurons that arise in the spinal cord and go on to innervate muscles and glands throughout the body. The interface between a motor neuron and muscle fiber is a specialized synapse called the *neuromuscular junction*, which is covered in greater detail later in this chapter. Interneurons are found only in the CNS and connect one neuron to another.

Specific glial cells found along the length of the neurons form myelin sheaths around the neurons. In the CNS, these are called *oligodendrocytes*, and in the PNS they are called *Schwann cells*. Surrounding the synapse space between neurons are astrocytes, which are specialized glial cells that deliver energy to neurons, among other functions (Sherwood et al. 2006). Astrocytes outnumber neurons by more than fivefold and make contact with both blood capillaries and neurons in the central nervous system to help form the blood–brain barrier, a structure that allows the blood vessels that deliver oxygen and nutrients to all of the tissues of the CNS to tightly regulate the movement of ions, molecules, and cells between the blood and the brain. This precise control of CNS homeostasis allows for proper neuronal function and also protects the neural tissue from toxins and pathogens. Alterations of these barrier properties are a significant component of pathology and progression of different neurological diseases (Daneman and Prat 2015).

CENTRAL NERVOUS SYSTEM

The CNS consists of the brain, the spinal cord, and the retinas of the eyes. The adult brain can be separated into four major regions: the cerebrum, the diencephalon, the brain stem, and the cerebellum.

Cerebrum

The cerebrum is the largest portion of the brain and contains the cerebral cortex and subcortical nuclei, collections of neurons that serve as the primary location for the production of the neurotransmitter acetylcholine. The cerebrum is made up of two hemispheres, while the cortex can be separated into four lobes: the frontal lobe, the parietal lobe, the temporal lobe, and the occipital lobe (figure 2.2). The frontal lobe is involved in motor function including facial expression, problem solving, spontaneity, memory, language, initiation, judgment, impulse control, and social and sexual

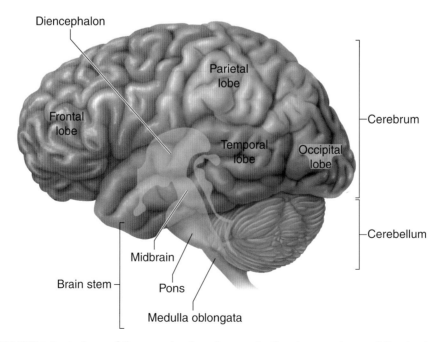

FIGURE 2.2 Lobes of the cerebral cortex and other key regions of the brain.

behavior. The most anterior portion of the frontal lobe is the prefrontal cortex, a critical portion that mediates intellectual or executive functions.

The other lobes are responsible for sensory functions. The parietal lobe processes information from the body surface and its interaction with the environment. This is referred to as *somatosensation*. The occipital lobe is where visual processing begins, and the temporal lobe contains the cortical area for auditory processing but also has regions crucial for memory formation.

Nestled deep within the temporal lobe are the two amygdalae. The amygdalae are almond-shaped regions composed of multiple clusters of neurons that receive information about the external environment from the sensory thalamus (more on this in a moment) and sensory cortices. The amygdalae are rapid detectors of aversive

Yoga and Neuroplasticity

Neuroplasticity describes the nervous system's ability to change in response to experience. So essentially the brain can direct its own changes. The roles of different parts of the cortex that we have discussed are not completely predetermined. In his book, *Livewired: The Inside Story of the Ever-Changing Brain*, David Eagleman (2020) describes how it is the pattern of inputs that determine the fate of the cortex. Eagleman goes on to explain how regions of the brain maintain their territory with continuous activity: If activity slows or stops (e.g., because of blindness), the territory tends to be taken over by its neighbors, a change that is measurable within an hour. Eagleman also hypothesizes that the circuitry underlying dreaming serves to amplify the visual system's activity periodically throughout the night, allowing it to defend its territory against takeover from other senses.

Although children, teenagers, and young adults have a greater capacity for neuroplasticity, we all have neuroplasticity throughout our lives. As adults, we can encourage neuroplasticity by seeking novel challenges and applying focused, deliberate effort. The neurotransmitter glutamate (not to be confused with the food additive monosodium glutamate or MSG) is the primary mediator of nervous system plasticity (Zhou and Danbolt 2014), and plenty of rest is required to lock in the changes that have been made. Perhaps this is what is meant when yogis say that Savasana (Corpse Pose) is a chance for the mind to imprint changes—but even better is sleep. A review by Walker and Stickgold (2004) provided evidence of sleep-dependent memory consolidation and sleep-dependent brain plasticity. Therefore, getting adequate amounts of sleep can help us retain new information, and inadequate sleep can slow memory formation.

In a randomized controlled trial, Tolahunase and colleagues (2018) looked at the effect that yoga and meditation can have on people with major depressive disorder. The authors concluded that a decrease in depression severity following yoga and meditation interventions is associated with improved systemic biomarkers of neuroplasticity. In a systematic review by Gothe and colleagues (2019) looking at the effect yoga can have on brain health, the authors concluded that yoga has a positive effect on the structure and function of various regions of the brain, including the amygdalae and prefrontal cortex. Gothe and colleagues suggested that these studies offer promising early evidence that behavioral interventions like yoga may mitigate age-related and neurodegenerative declines as many of the regions identified are known to demonstrate significant age-related atrophy.

environmental stimuli and situations, producing affective or behavioral states to allow for adaptive responses to potential threats. The amygdalae have long been associated with emotion and motivation, playing an essential part in processing both fearful and rewarding environmental stimuli, and they are implicated in a wide range of conditions, including addiction, autism, and anxiety disorders (Janak and Tye 2015). In addition to its role in emotion, the amygdala is also involved in the regulation or modulation of a variety of cognitive functions, such as attention, perception, and explicit memory.

Diencephalon

The diencephalon can be found just above the brain stem between the cerebral hemispheres and consists of four parts: the thalamus, the hypothalamus, the epithalamus, and the subthalamus. The thalamus is a relay between the cerebrum and the rest of the nervous system and plays an important role in regulating states of sleep and wakefulness. The hypothalamus coordinates homeostatic functions by linking the nervous system to the endocrine system via the pituitary gland, which is connected to it (figure 2.3). We explore the relationship between the hypothalamus and the pituitary gland more later in the chapter. The epithalamus consists primarily of the pineal gland—a tiny endocrine gland that secretes the hormone melatonin, which plays an important role in the regulation of our circadian rhythms or internal body clock (physical, mental, and behavioral changes that follow a daily cycle). The subthalamus is involved with the integration of skeletal muscle movements.

Brain Stem

The brain stem is the lower region of the brain and is structurally continuous with the spinal cord. It comprises the midbrain, the pons, and the medulla oblongata (figure

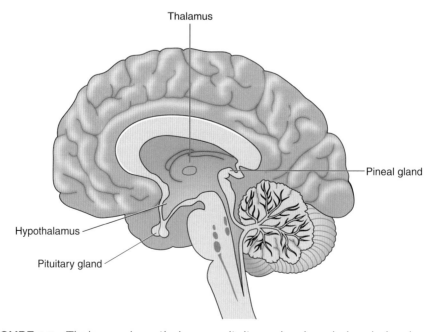

FIGURE 2.3   Thalamus, hypothalamus, pituitary gland, and pineal gland.

2.2). The brain stem plays an important role in the regulation of cardiac and respiratory function, consciousness, and the sleep cycle, as well as regulating vasodilation (the widening of blood vessels) and reflexes such as vomiting, coughing, sneezing, and swallowing. Connections from the motor and sensory systems of the cortex also pass through the brain stem to communicate with the PNS. The brain stem provides the main motor and sensory innervation to the face and neck via the cranial nerves.

Cerebellum

The cerebellum is connected to the brain stem, primarily at the pons, and is responsible for fine tuning voluntary movements and balance. The cerebellum maintains posture, controls muscle tone, and regulates voluntary muscle activity, therefore coordinating gait, but it is unable to initiate muscle contraction. Our procedural memories (often referred to as *muscle memory*) are formed in the cerebellum, which embodies more than two-thirds of all neurons in our brain (De Zeeuw and Ten Brinke 2015).

Spinal Cord

The spinal cord is a long, thin, tubular bundle of nervous tissue that extends from the medulla oblongata of the brain stem and runs along the inside of the vertebral column to the lumbar region. It serves as the signaling conduit between the brain and the periphery. Thirty-one pairs of spinal nerves branch from the spinal cord and form part of the PNS.

The Emotional Brain

The *limbic system*, while not a major organ system like the digestive system, is a convenient term to describe several functionally and anatomically interconnected structures found on the central underside of the cerebrum, comprising inner sections of the temporal lobes and the bottom of the frontal lobe. The limbic system combines higher mental functions and primitive emotion into a single system often referred to as the *emotional brain*. It is not only responsible for our emotional lives but also our higher mental functions, such as learning and formation of memories. The primary structures within the limbic system include the amygdala, hippocampus, thalamus, hypothalamus, and cingulate gyrus. The hippocampus is a complex brain structure embedded deep into the temporal lobe that has a major role in learning and memory. The cingulate cortex is located in the medial walls of the cerebral hemispheres and is an important interface between emotional regulation, sensing, and action.

In the book *A General Theory of Love*, three psychiatrists introduced the term *limbic resonance*, hoping to explain what we know about love from a scientific and neurological point of view (Lewis, Amini, and Lannon 2007). The theories presented in this book describe how people can literally be on the same wavelength and how this happens in the brain's limbic system.

Retinas

During embryonic development, the retina and optic nerve extend from the diencephalon and are thus considered part of the CNS. The retina is composed of layers of specialized neurons that are interconnected through synapses. As an extension of the CNS, the retina displays similarities to the brain and spinal cord in terms of anatomy, functionality, response to insult, and immunology. Several major neurodegenerative disorders have manifestations in the retina, suggesting that the eye is a window into the brain (London, Benhar, and Schwartz 2013).

Meninges, Ventricles, and Cerebrospinal Fluid

Three mater layers called the *meninges* encase the brain and spinal cord. From superficial to deep, these layers are the dura mater, arachnoid mater, and pia mater. The dura mater is a dense connective tissue layer that is connected to the inner surface of the skull. Next is the arachnoid mater, which is a thin, impermeable layer, and the innermost layer is the pia mater, which is a vascular layer that closely invests over the brain and spinal cord. Meningitis is a rare and potentially devastating infection that affects the delicate meninges. The meninges define three potential spaces: the epidural space, which exists between the skull and the dura mater; the subdural space, found between the dura mater and arachnoid mater; and the subarachnoid space, which is between the arachnoid mater and pia mater.

The cerebral ventricles are a series of interconnected, fluid-filled spaces that lie in the core of the forebrain and brain stem. The ventricles produce cerebrospinal fluid (CSF) and transport it around the cranial cavity via the subarachnoid space. CSF is an ultrafiltrate of blood plasma that performs vital functions, including providing nourishment, waste removal, and protection to the brain (Spector, Robert Snodgrass, and Johanson 2015). CSF acts as a shock absorber, cushioning the brain against the skull, and allows the brain and spinal cord to become buoyant, drastically reducing the effective weight of the brain and therefore the force applied to the brain and cerebral vessels during mechanical injury. The blood–CSF barrier also serves to regulate the environment of the brain. Adult CSF volume is estimated to be 150 ml, with a distribution of 125 ml within the subarachnoid spaces and 25 ml within the ventricles. CSF is completely renewed four to five times per 24-hour period in the average young adult (Sakka, Coll, and Chazal 2011).

PERIPHERAL NERVOUS SYSTEM

The peripheral nervous system (PNS) consists of all the neurons outside the brain and spinal cord. Bundles of axons in the PNS are referred to as *nerves*. Nerves are composed of more than just nervous tissue; they have connective tissues invested in their structure, as well as blood vessels supplying the tissues with nourishment. Each individual axon is surrounded by loose connective tissue, and many axons are then grouped together into fascicles, which are each surrounded by their own layer of fibrous connective tissue. Finally, multiple fascicles are grouped together to form a nerve, which is surrounded by its own layer of fibrous connective tissue. These three layers are similar to the connective tissue sheaths for muscles.

Yoga Inversions Bring More Blood to the Brain and Stimulate the Pineal Gland

There is widespread belief that inverting the body during asanas such as Headstand (Sirsasana) has many potential benefits including increasing blood flow to the brain and stimulating the pineal gland. However, the brain has the very important ability to maintain relatively constant blood flow despite changes occurring elsewhere in the body. In healthy adults, large changes in blood pressure result in little or no change in cerebral blood flow (Paulson, Strandgaard, and Edvinsson 1990). This mechanism of autoregulation of cerebral blood flow is vital since the brain is very sensitive to too much or too little blood flow. Only in severe head injury or acute ischemic stroke do we lose this autoregulation, leaving surviving brain tissue unprotected against the potentially harmful effect of blood pressure changes. So, it is reassuring to know that whether you regularly invert your body or not, your brain is receiving just the right amount of blood supply to meet its demands.

The pineal gland was once referred to as the *third eye* because of its location in the geometric center of our brain. The French philosopher, mathematician, and scientist Descartes regarded the pineal gland as the principal seat of the soul and the place in which all thoughts are formed. It is roughly the size of a soybean and is considered a somewhat mysterious organ because it was the last of the endocrine glands to have its function discovered. In addition to producing melatonin, the pineal gland also produces extremely tiny amounts of N,N-dimethyltryptamine, a potent psychedelic. While it has been proposed that the pineal gland excretes large quantities of N,N-dimethyltryptamine during extremely stressful life episodes, notably in the event of birth and death, to produce out-of-body experiences, there is a lack of evidence to back up this claim (Nichols 2018). This gland has the highest calcification rate among all organs and tissues of the human body. Pineal calcification is thought to jeopardize the amount of melatonin that can be produced by the gland and may be associated with a variety of neuronal diseases (Tan et al. 2018). There is, however, no research on the effect that yoga can have on the pineal gland. Maybe one day we will discover that Headstands decalcify the pineal gland, but until then just enjoy the literal change in perspective.

Nerves are associated with the region of the CNS to which they are connected, either as cranial nerves connected to the brain or spinal nerves connected to the spinal cord. There are 12 pairs of cranial nerves that exit the skull and are primarily responsible for the sensory and motor functions of the head and neck. One of these nerves targets organs in the thoracic and abdominal cavities. They can be classified as sensory nerves (afferent nerves), motor nerves (efferent nerves), or a combination of both.

There are 31 pairs of spinal nerves that exit along the length of the vertebral column and are named for the level at which each one emerges; for example, T1 is the thoracic nerve that exits from the spinal column from below the uppermost thoracic vertebra. The arrangement of these nerves is much more regular than that of the cranial nerves. All the spinal nerves are combined sensory and motor axons.

The sensory division of the PNS carries sensory information from the body to the central nervous system. The five senses we think of most are taste, smell, touch, sight, and hearing, which all receive stimuli from the outside world. Additional sensory stimuli come from within our internal environment. *Somatosensation*, often referred to as the *sixth sense*, is an all-encompassing term that includes the subcategories of mechanoreception (vibration, pressure, discriminatory touch), thermoreception (temperature), nociception (pain), equilibrioception (balance), and proprioception (sense of positioning and movement). Sherrington (1906) first described proprioception as "our ability to sense where our limbs and joints are in relation to our body and to the surrounding environment (position as well as movement) in the absence of visual feedback" (p. 17). Our sense of proprioception is fed by receptors found within our muscles, connective tissue, capsuloligamentous structures, and skin, which detect mechanical tissue changes and subsequently send sensory information for cerebral interpretation. A recent study by Cherup and colleagues (2020) concluded that a combination of yoga and meditation enhanced proprioception and balance in a group of individuals diagnosed with Parkinson's disease. Another recent study found significant improvements in balance and proprioception in amateur athletes following eight weeks of yoga practice (Sarhad Hasan, Haydary, and Gandomi 2020). A systematic review by Jeter and colleagues (2014) suggested that yoga may have an overall beneficial effect on balance; however, differences in quality of reporting and study design made it difficult for the authors to draw definitive conclusions.

The motor division of the PNS transfers signals from our central nervous system to our muscles, organs, and glands. The area where a neuron reaches a muscle fiber is called the *neuromuscular junction* (figure 2.4). As the nerve impulse arrives at the neuromuscular junction, it causes an influx of calcium ions into the neuron, which, in turn, releases the neurotransmitter acetylcholine. Once acetylcholine has passed across the synapse of the neuromuscular junction, sodium ions enter the muscle fiber, triggering an electrical charge (or action potential) that spreads throughout the muscle. The acetylcholine is eventually broken down, and the influx of sodium ceases, ending the action potential.

The PNS can be subdivided into the somatic nervous system and the autonomic nervous system.

Somatic Nervous System

The somatic nervous system (or voluntary nervous system) is the component of the peripheral nervous system associated with the voluntary control of body movements via skeletal muscles. So, during our yoga asana practice, we are constantly using our somatic nervous system. This system also transmits signals from receptors of external stimuli to the CNS, thereby mediating sight, hearing, and touch. The somatic nervous system therefore consists of both sensory (afferent) nerves and motor (efferent) nerves. The somatic nervous system also provides us with reflexes, which are automatic and do not require input or integration from the brain to perform. Monosynaptic reflexes, such as the knee-jerk reflex, have only a single synapse between the sensory neuron that receives the information and the motor neuron that responds. Polysynaptic reflexes have at least one interneuron between the sensory neuron and the motor neuron. An example of a polysynaptic reflex is seen when we step on something sharp; in response, our body must pull that foot up while simultaneously transferring balance to our other leg.

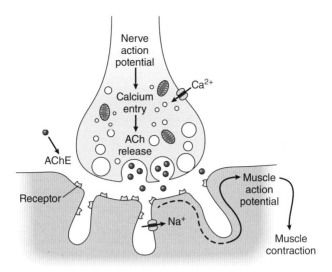

FIGURE 2.4 Neuromuscular junction.

Autonomic Nervous System

The autonomic nervous system is the division of the PNS that regulates involuntary physiologic processes including heart rate, blood pressure, respiration, digestion, and sexual arousal. This system has three divisions: the sympathetic nervous system (our fight-or-flight response); the parasympathetic nervous system (our rest-and-digest response); and the enteric nervous system (our second brain). Figure 2.5 compares the main features of the sympathetic nervous system and the parasympathetic nervous system.

Sympathetic Nervous System

The sympathetic nervous system (SNS) works with the endocrine system to trigger the *fight-or-flight response*, a term coined by Cannon (1915) to describe an animal's immediate response to danger. This is often referred to as the *stress response*, and while the word *stress* tends to have a negative connotation, a degree of stress is vital for our everyday functioning and survival. Without this system, it would be even harder to get out of bed in the morning, let alone run to catch a bus, or step out of the way of a runner on the sidewalk. The SNS is composed of many pathways that innervate nearly every living tissue in the body and is triggered, via the amygdalae, whenever we experience a situation that our brain perceives as being threatening. Danger, pain, upsetting feelings, and low blood sugar all activate the SNS. Muscular contractions are sympathetic in origin and therefore even a large component of our yoga asana practice is linked with the SNS.

The SNS primarily regulates blood vessels. An increase in sympathetic signals leads to vasodilation (widening) of the coronary vessels (vessels that supply the cardiac muscle of the heart) and the vessels that supply the skeletal muscles and external genitalia. All other vessels in the body will vasoconstrict (narrow). Sympathetic activation increases our heart rate, increases the contractile force of the heart, raises our blood pressure, decreases motility of the large intestine, and causes pupillary dilation and perspiration. The SNS also directly stimulates the adrenal glands to produce the hormone and

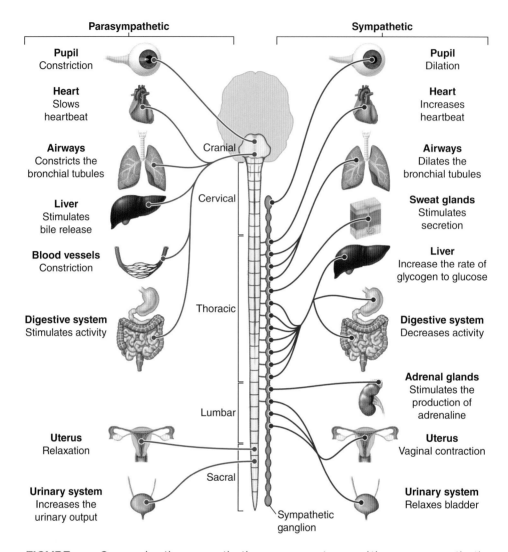

Parasympathetic

Pupil
Constriction

Heart
Slows
heartbeat

Airways
Constricts the
bronchial tubules

Liver
Stimulates
bile release

Blood vessels
Constriction

Digestive system
Stimulates activity

Uterus
Relaxation

Urinary system
Increases the
urinary output

Cranial

Cervical

Thoracic

Lumbar

Sacral

Sympathetic

Pupil
Dilation

Heart
Increases
heartbeat

Airways
Dilates the
bronchial tubules

Sweat glands
Stimulates
secretion

Liver
Increase the rate of
glycogen to glucose

Digestive system
Decreases activity

Adrenal glands
Stimulates the
production of
adrenaline

Uterus
Vaginal contraction

Urinary system
Relaxes bladder

Sympathetic
ganglion

FIGURE 2.5 Comparing the sympathetic nervous system and the parasympathetic nervous system.

neurotransmitter adrenaline (also known as epinephrine). All these actions prepare the body for immediate physical action. The SNS is constantly active even in nonstressful situations; for example, it is active during the normal respiratory cycle when sympathetic activation during inspiration dilates the airways, allowing for an appropriate inflow of air. The SNS also works to regulate your blood pressure every time you stand up.

The endocrine system is also heavily involved in the stress response. A diverse collection of neurons from multiple brain regions including the brain stem and amygdalae innervate specific neurons in the hypothalamus, which synthesizes and secretes a hormone called *corticotropin-releasing factor*. Corticotropin-releasing factor targets the neighboring pituitary gland, often referred to as the *master gland*, which, in turn, releases a substance called *adrenocorticotropic hormone*. Adrenocorticotropic

Is There a Difference Between Anxiety and Excitement?

It is important to note that feelings of excitement also stimulate the SNS. Our nervous systems cannot easily distinguish between anxiety and excitement, and it prepares the body to respond whether we are dealing with something truly dangerous or simply something new or unknown. Anxiety and excitement are both aroused emotions. The only difference is that excitement is a positive emotion often involving optimism.

The *anticipation* of something really wonderful or really bad can also stimulate the SNS, even if that anticipation is exaggerated or completely incorrect. In this sense, the brain does not distinguish between fantasy and reality. Think about the last time you had a nightmare. You will have woken up feeling panicked, flushed, breathing rapidly, and maybe even sweating. The SNS was triggered as if the nightmare was actually happening. The same goes with rumination: If we constantly worry that something might go wrong, we are triggering the SNS. This is where yoga and mindfulness practices play a key role. These practices help us to focus our attention so that our mind wanders less freely, and they help us to become more aware of thoughts. As we become aware of our thoughts, we come to the realization that we are not defined by them but are the observer of them. This sense of detachment from our thoughts can play a huge role in creating balance in our autonomic nervous system.

---

hormone consequently acts on the adrenal gland and triggers the release of cortisol. This release causes the acceleration of heart and lung action, constriction of blood vessels in many parts of the body, metabolism of fat and glucose for muscular action, dilation of the blood vessels supplying our major muscle groups, relaxation of the bladder, inhibition of erection, loss of hearing, loss of peripheral vision, and shaking. The systems that are not required during the stress response become temporarily suppressed, including the immune, digestive, and reproductive systems.

The actions of the endocrine system here are normally tightly regulated to ensure that the body can respond quickly to stressful events and return to a normal state just as rapidly. In the brain, cortisol participates in a negative feedback loop, meaning that it targets the hypothalamus to control its own production. This full cycle is known as the *hypothalamic-pituitary-adrenal* (HPA) *axis* (figure 2.6).

Barlow (2002) suggested that a freeze response can occur in some threatening situations. It is understood that freezing can be activated at intermediate levels of threat, when fleeing or aggressive responses are likely to be ineffective. In the context of predatory attack, some animals will freeze or play dead; this includes motor and vocal inhibition with an abrupt initiation and cessation (Schmidt et al. 2008). The freezing response is again initiated in the amygdalae (Applegate et al. 1983) and both the SNS and the parasympathetic nervous systems become activated (Iwata, Chida, and LeDoux 1987). The physiological parameters will vary, depending on which system is dominant at a certain point in time. Freezing is therefore not a passive state but can be thought of as a parasympathetic brake on the motor system—or attentive immobility.

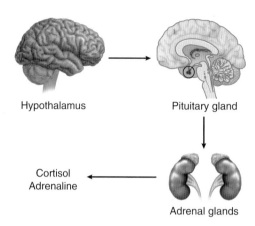

FIGURE 2.6 Hypothalamic-pituitary-adrenal (HPA) axis.

Parasympathetic Nervous System

The parasympathetic nervous system (PSNS) works in opposition to the SNS to balance, calm, and restore the body. Its effect is often referred to as the *rest-and-digest response*. The vagus nerve is the 10th cranial nerve. It makes up about 75 percent of the PSNS. It is the longest cranial nerve in the body, extending from the head to the abdomen (the word *vagus* means wandering in Latin), and it provides parasympathetic input to most of the abdominal organs (except the adrenal glands and the descending colon), the muscles of the throat, the soft palate, the larynx, and part of the outer ear (figure 2.7). The PNS can affect specific organs and is therefore not an all-or-nothing response (McCorry 2007). It causes decreased heart rate, vasodilation of blood vessels, decreased respiratory rate, increased motility in the digestive tract, and release of digestive enzymes and insulin from the pancreas. Around 80 percent of the fibers of the vagus nerve are afferent, carrying sensory information to the CNS from our organs regarding how safe or unsafe we feel, therefore helping us to regulate our stress response.

So, how can we stimulate the PSNS? The key to this begins with a natural phenomenon occurring as part of our physiology: Parasympathetic nerves fire during expiration, contracting and stiffening airways to prevent collapse. Therefore, by lengthening our exhalation, we can directly target our PSNS. This is also why our heart rate naturally decreases with every exhalation (and naturally increases with every inhalation due to SNS activation). Singing, playing wind instruments, chanting, pranayama, and yoga can all stimulate the PSNS in this way. A pilot study by Kalyani and colleagues (2011) concluded that chanting deactivated the limbic system, and the authors proposed that this was mediated by stimulation of the vagus nerve. Bernardi and colleagues (2001) found that recitation of the rosary prayers, and also of yoga mantras, slowed the respiratory rate to almost exactly six breaths per minute and enhanced parasympathetic activity. In addition to this, there is evidence that humming, through the action of lowering the pitch of your voice and creating resonance in the throat, can trigger the PSNS. A pilot study by Sujan and colleagues (2015) suggested that Bhramari pranayama (humming bee breath) increases parasympathetic activity.

There are many additional ways in which we can potentially stimulate the PSNS. When we are feeling stressed, our peripheral vision narrows so that we can focus on any imminent threats. By consciously focusing on our peripheral vision, we can potentially shift ourselves out of the stress response by triggering the PSNS. There is also preliminary evidence that acclimating to cold conditions lowers sympathetic activation and causes a shift toward increased parasympathetic activity (Mäkinen et al. 2008). This could be explored by regularly having cold showers. Research has even suggested that omega-3 fatty acids increase vagal tone and vagal activity (O'Keefe et al. 2006). A review by He and colleagues (2012) concluded that auricular (ear) acupuncture plays a role in increasing parasympathetic activity. Lu, Chen, and Kuo (2011) suggested that foot reflexology can increase vagal modulation, decrease sympathetic modulation, and lower blood pressure in healthy subjects and patients with coronary artery disease. Invited social engagement is also a key element. Porges (2011) has provided exciting insights into the way our autonomic nervous system unconsciously mediates social engagement, trust, and intimacy. Allowing ourselves to be truly vulnerable with others and accepting other people's vulnerabilities can help us

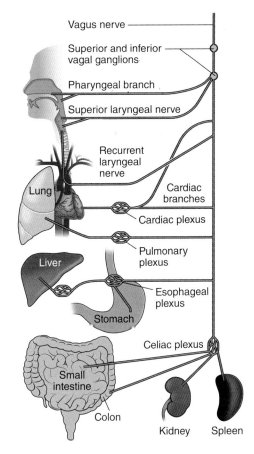

FIGURE 2.7 Branches of the vagus nerve.

TRY IT YOURSELF: Bhramari Pranayama

Bhramari pranayama, or bee breath, can be a very calming practice for the mind. Start by finding a comfortable seat on the floor, in a chair, or by propping yourself up in bed. Gently press your tragi, the pieces of cartilage between your cheeks and your ears, with your thumbs, and rest your other fingertips on your skull. Close your eyes if that feels accessible to you today; otherwise soften your gaze and focus on a fixed object in front of you. Release your jaw, let your lips and teeth gently part and allow your tongue to fall away from the roof of your mouth. Start by taking a couple of gentle breaths in and out through your nose, and as you next exhale, make the sound of the letter *M*, essentially a humming sound. Repeat this for six or seven rounds and then return to simple nasal breathing. Take a moment to notice how you feel. If you want to practice this pranayama in a public space and feel self-conscious, you can practice silent Bhramari for six or seven rounds; inhale and then exhale as you imagine you are humming without making the sound.

to feel safe and access our PSNS. A study by Kok and colleagues (2013) concluded that increased positive emotions, in turn, produced increases in vagal tone, an effect mediated by increased perceptions of social connections.

Enteric Nervous System

The enteric nervous system (ENS), sometimes referred to as the *second brain*, is embedded in the lining of the gastrointestinal system and is the largest component of the autonomic nervous system. When we talk about a gut feeling or "butterflies" in the stomach, we are essentially referring to the ENS. There is bidirectional information flow between the ENS and CNS. While the ENS normally communicates with the CNS via the PNS and SNS, it is composed of sensory neurons, motor neurons, and interneurons, all of which make it capable of carrying reflexes and acting as an integrating center in the absence of CNS input (Rao and Gershon 2016). The ENS has essential functions including controlling the motor functions of the gut (the tone of the gut and the velocity and intensity of its muscular contractions), local blood flow, mucosal transport, and secretions, and modulating immune and endocrine functions.

The human gastrointestinal tract harbors a complex and dynamic population of microorganisms known as the *gut microbiota*. The genetic content of the microbial communities in our gut outnumbers the human genetic content of our whole body by approximately a hundredfold (Ley, Peterson, and Gordon 2006). In a review by Carabotti and colleagues (2015), the authors concluded that strong evidence suggests that gut microbiota has an important role in bidirectional interactions between the gut and the nervous system. It interacts with the CNS by regulating brain chemistry and influencing the neuroendocrine systems associated with stress response, anxiety, and memory function.

While antibiotics are crucial for fighting many serious infections, their use has a huge impact on our gut microbiota and therefore on the relationship between the gut and the nervous system. A study by Lurie and colleagues (2015) reported that one course of antibiotics can increase the risk of major depression by 24 percent and anxiety disorders by 17 percent. The study also found that two courses of antibiotics in a one-year period increased the risk of major depression by 52 percent and anxiety disorders by 44 percent.

CONDITIONS OF THE NERVOUS SYSTEM

We will now explore some of the main conditions that affect the nervous system and discuss the potential role that yoga can play in improving these conditions.

Persistent Stress and HPA Axis Dysfunction

Earlier we discussed how vital the stress response is and how it is constantly active even in nonstressful situations. In an ideal world, we all have varying levels of stress, which provide us an opportunity to respond quickly and effectively to all the demands that we meet throughout each day. However, modern-day lifestyles tend to be challenging for most people's nervous systems. So many of us experience regular information overload thanks to all our digital devices, along with the feeling of always being switched on. We are often stimulated by artificial light, high frequency sounds, and air pollution. We are trying to manage extreme circumstances such as global pandemics, political unrest, and rapid climate change, to name but a few. A study by Almeida and colleagues (2020) examined how stress in the daily lives of Americans may have changed over the last few decades, finding a stark increase in day-to-day stress among adults, particularly from 45 to 64 years old, in the 2010s compared to the 1990s.

In 1998, chiropractor James Wilson coined the term *adrenal fatigue* and used it to describe a condition in which the adrenal glands, overstimulated by chronic stress, burn out and shut down, causing a variety of symptoms. The concept of adrenal fatigue is that cortisol production is constantly triggered, which eventually wears out the adrenal glands. However, many people who believe they have adrenal fatigue often do not have dysfunctional cortisol levels. A systematic review by Cadegiani and Kater (2016) reported that there is no substantiation that adrenal fatigue is a physiological reality. However, HPA axis dysfunction, which is an alteration in stress response over time (after exposure to chronic stress), has been verified and associated with numerous diseases. Long-term SNS activation has been linked to type 2 diabetes, obesity, and cardiovascular disease (Chrousos 2009). Cortisol has also been demonstrated to have detrimental effects on memory and cognition (Newcomer et al. 1999), and high cortisol levels are implicated in mood disorders like depression (Moylan et al. 2013). Additionally, baseline sympathetic activity can be affected by early life experiences, and some studies suggest that early life trauma may lead to an overreactive HPA axis later in life (Liu et al. 2000). This may contribute to increased anxiety and potential metabolic effects, including excess fat deposition and insulin resistance (Maniam, Antoniadis, and Morris 2014). Chronic stress exposure has also been found to lead to psychological disturbances such as depression (Charney and Manji 2004). Earlier in the chapter we discussed how the immune, digestive, and reproductive systems are temporarily suppressed during the typical stress response. When the stress response becomes prolonged, this can lead to fertility issues, persistent problems with digestion, and weakened immune responses.

Can Yoga Help to Relieve Persistent Stress and Anxiety?

A review by Li and Goldsmith (2012) looked at the effects of yoga on anxiety and stress. The review included 35 studies and concluded that while yoga can relieve stress and anxiety, further investigation into this relationship using large, well-defined populations, adequate controls, randomization, and long duration should be explored before recommending yoga as a treatment option. In a systematic review of randomized controlled trials on the effects of yoga on stress measures and mood, the 25 studies that were included provided preliminary evidence to suggest that yoga leads to better regulation of the SNS and the HPA axis, as well as a decrease in depressive and anxious symptoms in a range of populations (Pascoe and Bauer 2015). A systematic review by Sharma and Haider (2013) focusing on the effect that yoga has on anxiety concluded that out of a total of 27 studies that met their inclusion criteria, 19 studies demonstrated a significant reduction in anxiety.

A review on neuroimaging in yoga practitioners showed decreased blood flow in the amygdala and increased activity in the prefrontal cortex, suggesting that practitioners do notice negative stimuli but are less affected by it (Desai, Tailor, and Bhatt 2015). Partaking in meditation and yoga practice is associated with smaller right amygdala volume (Gotink et al. 2018) and stress reduction has been associated with less amygdala volume (Holzel et al. 2010). Streeter and colleagues (2010) suggested that yoga increases PNS activity and neurotransmitter (GABA) levels in the thalamus, and that these increases are correlated with reduced anxiety and improved mood.

Depression

Depressive disorders, such as major depressive disorder, are the leading cause of disability worldwide, affecting more than 340 million people (Greden 2001). The World Health Organization (2012) projects that depression will be the world's leading disease by 2030.

A systematic review and meta-analysis by Cramer, Lauche, Langhorst, and Dobos (2013) concluded that yoga could be considered a supplementary treatment option for patients with depressive disorders and individuals with elevated levels of depression. A recent systematic review of 19 studies and a meta-analysis of 13 studies (Brinsley et al. 2020) looked at the effects of yoga on depressive symptoms in people with mental disorders including depression, posttraumatic stress, schizophrenia, anxiety, alcohol dependence, and bipolar disorder. The authors concluded that yoga showed greater reductions in depressive symptoms than the control groups, and greater reductions in depressive symptoms were associated with higher frequency of yoga sessions per week.

Some research has been conducted to explore whether gratitude can help to combat symptoms of depression. Watkins and colleagues (2003) suggested that grateful individuals have these four characteristics: They do not feel deprived in life, they appreciate others' contributions to their well-being, they tend to appreciate simple pleasures that are freely available to most people, and they acknowledge the important role of experiencing and expressing gratitude.

A literature review by Wood, Froh, and Geraghty (2010) reported that gratitude significantly lowered the risk of a range of diagnoses including major depression, generalized anxiety disorder, phobia, nicotine dependence, alcohol dependence, drug abuse or dependence, and the risk of bulimia nervosa. A more recent review by Jans-Beken and colleagues (2019) concluded that having a grateful disposition is positively linked to the absence of psychopathology, but gratitude interventions are not unequivocally established as universally effective for decreasing psychopathological symptoms.

TRY IT YOURSELF: Gratitude Meditation

Gratitude meditation is a wonderful practice to do at the start and the end of each day. Find any comfortable position and either gently close your eyes or simply soften your gaze and focus on a fixed point in front of you. Release your jaw, let your lips and teeth gently part and allow your tongue to fall away from the roof of your mouth. Start by taking a couple of gentle breaths in and out through your nose. Notice your abdomen gently expand and contract with each inhalation and exhalation. When you are ready, start to consider all the things we have today that make our lives easier and more comfortable than they were for our parents and grandparents. Take a few moments to reflect on the hundreds of people who have worked hard to make your everyday life easier or more pleasant. Then think about your family, friends, neighbors, and colleagues who enrich your life and support you. Consider your own reasons for feeling grateful in this moment. When you are ready, bring your focus back to your breath. Come back into your body by gently wiggling your fingers and toes. Notice how you are feeling after this short meditation.

Posttraumatic Stress Disorder

Posttraumatic stress disorder, often referred to as PTSD, is a condition that can develop after exposure to a traumatic event and is characterized by four hallmark clusters of symptoms lasting for more than a month: reexperiencing, avoidance, negative cognitions or mood, and hyperarousal (American Psychiatric Association 2013). While half of the cases resolve within three months, some people experience symptoms for extended periods or experience symptoms that resolve and reappear over time.

A systematic review and meta-analysis by Cramer and colleagues (2018) included seven trials and concluded that only a weak recommendation for yoga as an adjunctive intervention for posttraumatic stress disorder can be made. A systematic review and meta-analysis by Hilton, Ruelaz Maher, and colleagues (2017) included 10 trials and concluded that meditation appears to be effective for posttraumatic stress disorder and depression symptoms, but in order to increase confidence in findings, more high-quality studies are needed. It is worth noting that the potential risks of meditation, including for trauma survivors, have become increasingly well known. A study by Lindahl and colleagues (2017) examined the range of challenging experiences that can arise in the context of Buddhist meditation—experiences that can resemble psychological dissociation, depersonalization, and the reexperiencing of traumatic memories. Thankfully, with this knowledge has come the development of trauma-sensitive mindfulness and meditation practices.

Cognitive Impairment

Dementia and mild cognitive impairment (MCI) are characterized by decline from a previously attained cognitive level, but in dementia, as opposed to MCI, the decline impacts on activities of daily living or social functioning (World Health Organization 2016). In MCI, although one can still engage in complex activities—for example, paying bills or taking medication—greater effort or new strategies may be required. Dementia is usually preceded by MCI, and the boundary between the two is gray. There are many different causes of dementia, with Alzheimer's disease and *vascular dementia* (a general term describing problems with reasoning, planning, judgment, memory, and other thought processes caused by brain damage from impaired blood flow to the brain) being the most common. Globally, there were around 47 million people living with dementia in 2015, and this is projected to increase to 66 million by 2030 and 115 million by 2050 (Prince et al. 2015).

The impact of yoga on cognition is evident in a meta-analysis that reported moderate changes in attention, processing speed, and executive function measures for studies conducted with adult populations (Gothe and McAuley 2015). In 2017, Du and Wei conducted a systematic review of the therapeutic effects of yoga in individuals with dementia but were only able to include two studies in their review. The authors found that yoga significantly improved cognitive and motor functions and behavioral issues. In a review by Brenes and colleagues (2019) looking at the effects of yoga on patients with MCI and dementia, the authors included four studies and suggested that yoga may have beneficial effects on cognitive functioning, particularly on attention and verbal memory. Furthermore, yoga may affect cognitive functioning through improved sleep, mood, and neural connectivity. In a review in 2020, Bougea warned that evidence on the effects of yoga in patients with dementia is limited and conflicting in some cases. The review also highlights the need for longitudinal studies with a rigorous methodology to define the optimal frequency and intervals of yoga and to evaluate the cost-effectiveness for people with dementia. A systematic review by Green and colleagues (2019) reported that there is moderate evidence to support the use of yoga to decrease the risk for falls for community-dwelling older adults and people with dementia and Alzheimer's disease.

Parkinson's Disease

Parkinson's disease (PD) is a neurodegenerative disorder that affects predominantly dopamine-producing neurons in an area of the midbrain called the *substantia nigra*. Symptoms generally develop and progress slowly over years and may include slowness of movement, tremor, limb rigidity, and gait and balance problems. The cause of PD essentially remains unknown. PD affects 1 or 2 per 1,000 of the population at any time. The prevalence increases with age with 1 percent of the population above 60 years of age being affected (Tysnes and Storstein 2017).

A systematic review and meta-analysis by Jin and colleagues (2019) included 21 studies and concluded that mind–body exercises (including yoga) were found to lead to significant improvements in motor function, depressive symptoms, and quality of life in patients with PD, and they can be used as an effective method for clinical exercise intervention in PD patients. A systematic review by Green and colleagues (2019) examined the efficacy of yoga as a neuromuscular intervention for community-dwelling populations at risk for falls to determine its utility for use in occupational therapy intervention. The authors reported that studies involving people with PD did not include strong enough evidence to be able to make a clear classification. A more recent study by Cherup and colleagues (2020) concluded that a combination of yoga and meditation enhanced proprioception and balance in a group of individuals diagnosed with PD.

Persistent Pain

The International Association for the Study of Pain defines pain as "an unpleasant sensory and emotional experience associated with actual or potential tissue damage, or described in terms of such damage" (Merskey and Bogduk 1994, p. 209) and defines persistent pain as pain on most days or every day in the past six months. It is important to note that pain and tissue damage do not always correlate, and the longer a person experiences pain, the weaker the correlation. It is very possible to have tissue damage without experiencing pain and it is also possible to experience pain when there is no apparent tissue damage (Crofford 2015). Nearly every condition we assume to be a sign of dysfunction (poor posture, tightness, weakness, degeneration) can exist in people without the presence of pain. Pain might tell us that there is a problem, but it rarely tells us what the problem is, where it is, or how bad it is. Sadly, population-based estimates of persistent pain among U.S. adults range from 11 to 40 percent, meaning that a lot of people are in pain for a lot of time (Interagency Pain Research Coordinating Committee 2016).

Persistent pain is often correlated with depression, fear, rumination, and worries about injury. Recent studies have provided incontrovertible evidence that psychiatric disorders and other psychosocial factors can influence both the development of persistent pain conditions and the response to treatment. In a study by Polatin and colleagues (1993), 77 percent of patients with persistent low back pain met lifetime criteria, and 59 percent demonstrated current symptoms for at least one psychiatric diagnosis, with the most common being depression, substance abuse, and anxiety disorders. Notably, more than 50 percent of those with depression and more than 90 percent of patients with substance abuse or an anxiety disorder experienced symptoms from these psychiatric disorders before the onset of low back pain. Most, but not all, studies have shown untreated psychopathology to negatively affect low back pain treatment outcomes (Fayad et al. 2004). This does not mean that pain is all in one's head. It means that there is often a strong link between experiencing pain and one's mental resilience. Neugebauer and colleagues (2004) reported that neuroplastic changes were shown in the amygdalae in persistent pain. It is thought that the amygdalae play an important role in the emotional–effective dimension of pain (Neugebauer 2015).

Social factors have also been demonstrated to have an impact on persistent pain. These include return-to-work issues, catastrophizing, poor role models, codependency behavior, inadequate coping mechanisms, and attitudes, beliefs, and expectations (Seres 2003). Catastrophizing is a cognitive process whereby a person exhibits an exaggerated notion of negativity, assuming the worst outcomes and interpreting even minor problems as major calamities (Biggs, Meulders, and Vlaeyen 2016). Pain could be thought of as all of life's stressors (physical, social, psychological, and spiritual) exceeding our perceived ability to withstand or adapt to these. To optimize outcomes, the identification and treatment of associated psychosocial issues is of paramount importance. This is best accomplished via a multidisciplinary approach and using the biopsychosocial model, which was first presented by Engel in 1977. The biopsychosocial model (figure 2.8) is a way of understanding how suffering, disease, and illness are affected by multiple levels of organization, from the societal to the molecular. At the practical level, it is a

Can Yoga Help in the Management of Persistent Pain?

A systematic review and meta-analysis of mindfulness meditation for persistent pain by Hilton, Hempel, and colleagues (2017) concluded that there was low-quality evidence that mindfulness meditation is associated with a small decrease in pain compared with all types of controls in 30 trials. Statistically significant effects were also found for depression symptoms and quality of life. The authors recommended that additional well-designed, rigorous, and large-scale trials are needed to decisively provide estimates of the efficacy of mindfulness meditation for chronic pain. A systematic review and meta-analysis of yoga for low back pain by Cramer, Lauche, Haller, and Dobos (2013) reported that there was strong evidence for short-term effectiveness and moderate evidence for long-term effectiveness of yoga for chronic low back pain in the most important patient-centered outcomes. A systematic review of randomized controlled trials looking at the effects of yoga on chronic neck pain (Kim 2016) concluded that there was evidence from the three trials showing that yoga may be beneficial for chronic neck pain.

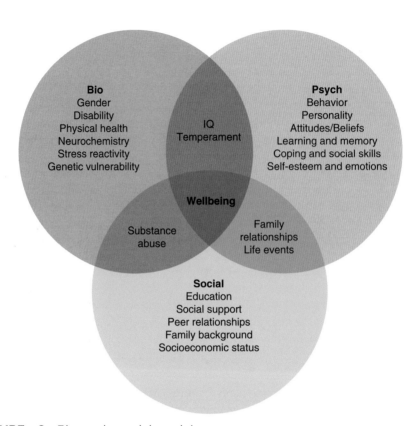

FIGURE 2.8   Biopsychosocial model.

way of understanding the patient's subjective experience as an essential contributor to accurate diagnosis, health outcomes, and humane care (Borrell-Carrió, Suchman, and Epstein 2004).

It is important for us all to realize that experiencing some pain at different stages in our lives is a normal part of being human. It is our response to the pain that tends to be most significant. Often understanding and acknowledging pain can be desensitizing.

CONCLUSION

The human nervous system is complex and fascinating, readily changing, and adapting in response to our experiences. Yoga is an important tool for helping us to create better regulation of our SNS and has the wonderful ability to help us to manage persistent pain, reduce depressive symptoms, and improve our cognition.

Chapter 3

RESPIRATORY SYSTEM

Yoga is rarely described without some mention of breath, and asana without breath awareness is just stretching. While modern postural yoga as we know it has only materialized over the last hundred years or so (Singleton 2010), the earliest mentions of yoga, which can be found in the *Rig Veda* written around 1500 BCE, describe it as a practice of meditation and breath control, or pranayama. The ancient yogis considered the breath a gateway to liberation.

Throughout history and across many cultures, breath has been synonymous with life force, and many cultures have language linking the two. The word *spirit* comes from the Latin *spiritus*, which means breath. In yogic and Hindu philosophy, *prana*, or vital life force, is said to be carried on the breath. The Polynesian *mana*, the Hebrew *ruach*, and the Greek *psyche* (as in psychology) are all related to the concept of breath as life force or soul.

To this day, breathing is seen not just as an exchange of gases but as the very essence of life and, arguably, the seat of our spirit. So, perhaps understanding our breath, including the anatomy and physiology of it, can help us understand our spirit.

Breathing can have a profound impact on all the other systems in the body, and breathing is a process that can be either involuntary, as it is much of the time, or voluntary, as it often is during a yoga practice. Practicing breath awareness and slowing of the breath can have profound implications on both our physiological and psychological well-being.

ANATOMY AND PHYSIOLOGY OF THE RESPIRATORY SYSTEM

The respiratory system (figure 3.1) is composed of the organs and other body parts involved in the process of breathing, or pulmonary ventilation. These organs include the nose and nasal cavity, sinuses, mouth, throat (pharynx), voice box (larynx), windpipe (trachea), diaphragm, lungs, bronchial tubes, air sacs (alveoli), and capillaries. The primary organs of the respiratory system are, of course, the lungs, which conduct the chief role of the whole system: to absorb oxygen and release carbon dioxide.

A Personal Note From Matt

When my father passed away in 2007, I was devastated. After contracting MRSA (methicillin-resistant *Staphylococcus aureus*) during a routine bunion surgery, he died at the age of 60, much too young in my opinion. Being grief stricken by his death, I thought the viewing of his body would be very difficult. To my surprise, I did not feel much upon seeing the body my father had inhabited for 60 years. Even though I had hugged that body just a week prior, I felt in me that it was no longer him. Indeed, with his last breath, also went his spirit. This realization made his death easier for me to accept, and, in a way, the breath I was breathing was the same as what he had breathed. If, as the many languages mentioned previously suggest, the spirit and the breath are one, then we are all indeed part of one spirit.

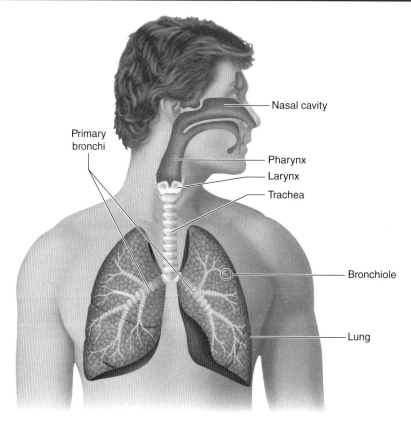

FIGURE 3.1   The respiratory system.

As we breathe in, air from the atmosphere enters the alveoli sacs; imagine millions of tiny, microscopic balloons. Capillary beds cover each tiny alveolus (figure 3.2) so that the blood can quickly absorb oxygen and release carbon dioxide, which is then expelled into the environment. Because the air within the alveoli has more oxygen than the blood, the oxygen diffuses across the alveoli to the blood while carbon dioxide diffuses out. Iron-rich hemoglobin in the blood can bind with both oxygen and carbon dioxide for transport.

Once oxygen is received via the lungs, the circulatory system pumps the oxygen-rich blood to all the cells of the body. The blood then collects carbon dioxide and other waste products, which are byproducts of cellular respiration, and pumps those to the lungs to be expelled.

Our cells need oxygen. They combine glucose and oxygen, which readily reacts with other substances, to produce energy in a process called *aerobic cellular respiration*. This process releases carbon dioxide, water, and, most importantly, adenosine triphosphate, the energy currency used by cells. We obtain oxygen through the process of breathing, which is also

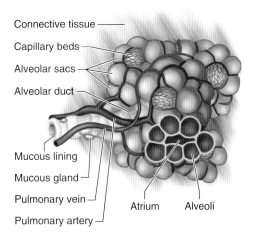

FIGURE 3.2 Alveolar sacs covered with capillary beds.

known as *respiration* but is more accurately called *pulmonary ventilation* to distinguish it from the respiration that happens at the cellular level. Some animals, like earthworms and frogs, obtain oxygen through their skin. However, we mammals need more oxygen because of our greater metabolic demands, so we draw in much more oxygen using our strong diaphragms and relatively large lungs.

THE BIOMECHANICS OF QUIET BREATHING

The lungs cannot inflate themselves, so we use pressure changes to draw in air. Boyle's law states that the pressure of a gas is inversely proportional to the volume of the container holding the gas. In simpler terms, this means that as the size of a closed container increases, the air pressure inside the container decreases. You can try this volume-to-pressure change yourself. After an exhale, close your mouth and pinch your nose so that no air can enter. Then, try to inhale. You will be increasing the size of your thoracic container and thus lowering the pressure of the air in your lungs. As you unpinch your nose, atmospheric air rushes in, following the pressure gradient you created, and you will make a gasp, sounding like the hydraulic brakes you might hear from a bus.

This increase in thoracic volume happens through the contraction of the diaphragm, the primary muscle of breathing. As the dome-shaped diaphragm subtly flattens, the chest cavity increases, which, following Boyle's law, lowers the pressure in the lungs, and air subsequently moves from a place of higher pressure (the environment) to a place of lower pressure (the lungs), thus creating an inhalation or inspiration. In the absence of dysfunctional breathing, which is discussed later, all quiet, normal breathing occurs because of the contraction of the diaphragm (figure 3.3). The intercostal muscles, which are located between the ribs, act in concert with the scalene muscles to expand the upper rib cage and to prevent it from being drawn inward by the action of the diaphragm, a phenomenon known as *paradoxical breathing* (Han et al. 1993). In other words, while the diaphragm is the main pump of breathing, other rib cage muscles help to stabilize the ribs against the pull created by the diaphragm.

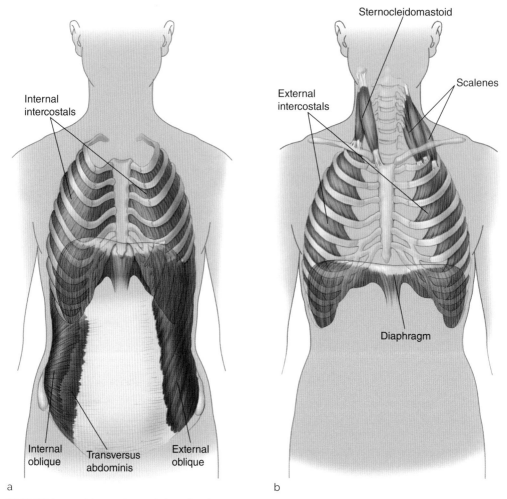

FIGURE 3.3 Excursion of the diaphragm: *(a)* the diaphragm in its relaxed dome shape; *(b)* the diaphragm at the bottom of its excursion and thus the top of the inhalation.

Coming from the Greek *diaphragma* for partition, the respiratory diaphragm separates the thorax (with the heart and the lungs) from the abdomen (with the abdominal viscera). As the diaphragm contracts, it presses down onto the abdominal contents, displacing them, and we inhale. It is important to note that when the diaphragm moves, the chest cavity experiences a shape change and a volume change, but the abdomen, being compressed downward, only experiences a shape change because it is effectively a sealed container.

At rest, exhalation—or expiration—is a passive process where the diaphragm and intercostal muscles relax, and the thorax shrinks with elastic recoil, decreasing thoracic volume and therefore increasing the air pressure. The air is then forced from the lungs back into the atmosphere. At the end of a passive exhale, the chest and abdomen will be in a resting position, which is determined by their anatomical elasticity. At this point, the lungs still contain some air, called the *functional residual capacity*, which in an adult is about 2.5 to 3.0 liters (figure 3.4).

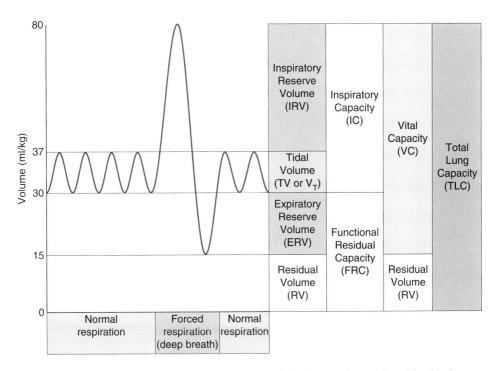

FIGURE 3.4 Normal ventilation, functional residual capacity, and residual volume.

Given that normal inhalation is active and exhalation passive, perhaps this is why in yoga we tend to inhale into more active movements, like backbending, and exhale into more passive movements, like softening into a forward fold.

ALL ABOUT THE DIAPHRAGM

The diaphragm is a very thin (2-4 mm), dome-shaped sheet of skeletal muscle and tendon that divides the torso in two. The diaphragm is slightly lower on the left to accommodate the heart and the two are connected through the heart's pericardium. During diaphragm excursion, the heart travels with it. You might be able to feel the descending movement of your heart as you breathe by lying down, taking slow, deep belly breaths, and placing your hand on your heart. If you pay close attention to where you feel the apex of your heartbeat, you might feel it move down as you inhale and up as you exhale.

The diaphragm has three openings: one for the inferior vena cava, which delivers deoxygenated blood to the heart; one for the aorta, which delivers oxygenated blood to the lower half of the body; and one for the esophagus, which moves food to the stomach, and the vagus nerve, which regulates digestion and receives information from the gut.

Diaphragmatic Connections to Other Parts

The diaphragm has many anatomic connections to other structures. It is linked by fascia with the abdominal muscles including the transverse abdominis, the pelvic floor, the psoas muscles, and the quadratus lumborum, to name but a few. Furthermore, the fascia to which the thoracic diaphragm connects can be seen to go from the cervical spine to the pelvic floor (Bordoni and Zanier 2013).

Indeed, the diaphragm is not a disconnected muscle working alone. Bordoni and Zanier (2013) conclude:

> The diaphragm muscle not only plays a role in respiration but also has many roles affecting the health of the body . . . The diaphragm muscle should not be seen as a segment but as part of a body system. To arrive at correct therapeutic strategies, we must see the whole and all the links . . . (p. 288).

The Diaphragm as a Core Muscle

In addition to being the primary pump in pulmonary ventilation, the diaphragm performs other important roles as well. Owing to its position and connection with other structures, the diaphragm aids with spinal stabilization.

Though there is no singular agreed definition of the *core*, not even among movement professionals or doctors, most of us tend to think of the core as the musculature supporting the abdominal organs and the spine. Surrounding the abdominal organs is a cylinder of musculature with the transverse abdominis wrapping around the sides of the abdomen and the diaphragm being the roof (figure 3.5). Creating the bottom of this cylinder is the pelvic floor, which also plays an important role in stabilizing the abdominal viscera. As the transverse abdominis engages to brace our midsection and spine before any full-body movement, the top of the cylinder (the diaphragm) and the bottom of the cylinder (the pelvic floor) also need to engage. In detailing lumbar stabilization, Barr, Griggs, and Cadby say, "As the roof of the cylinder of muscles that surround the spine and assist with stability, the diaphragm is a major contributor to intra-abdominal pressure and therefore lumbar stability" (2005, p. 476). Interestingly, at rest, the diaphragm and the pelvic floor move in a symmetrical way, as confirmed with real-time magnetic resonance studies on living subjects (Talasz et al. 2011).

Intra-abdominal pressure (IAP) is defined as the steady-state pressure concealed within the abdominal cavity resulting from the interaction between the abdominal wall and viscera (Milanesi and Caregnato 2016). The degree of IAP varies according to our inhalation, exhalation, and abdominal wall resistance. While elevated IAP can be the result of an acute abdominal syndrome and so a concern for intensive care units, within the context of a healthy individual, voluntarily created IAP through the engagement of the core musculature can be very important for stabilizing the torso and spine during full-body movements. During a yoga class, your abdomen will vary from high levels of IAP during such challenging movements as core work, handstands, and even balancing poses to very low levels of IAP during the restorative elements of the practice such as a seated forward fold or Savasana.

IAP is very important in helping to stabilize the spine and is created through the concerted contraction of what is sometimes called the *abdominal cylinder*, the

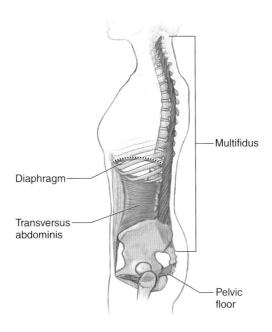

Multifidus

Diaphragm

Transversus
abdominis

Pelvic
floor

FIGURE 3.5 When engaged, the diaphragm, transverse abdominis muscles, and pelvic floor create a cylinder of intra-abdominal pressure that stabilizes our center and low back.

muscles of the abdominal wall, including the transverse abdominis and the obliques, the pelvic floor on the bottom, and the thoracic diaphragm on the top. It is believed that the transverse abdominal muscles are among the first muscles to activate to build IAP (Cresswell, Oddsson, and Thorstensson 1994) and for the transverse abdominis to increase tension in the thoracolumbar fascia, the diaphragm must also activate to prevent the abdominal contents from displacing (Hodges 2004). Furthermore, before any full-body movement, our diaphragm will contract to create abdominal pressure, which helps to stabilize the lumbar segment of the spine (Ebenbichler et al. 2001).

FORCED EXHALATION AND INHALATION

During exercise, the metabolic needs of the body are greater, and a quicker exchange of oxygen and carbon dioxide is necessary. If normal breathing cannot suffice, the body uses accessory muscles to speed up the process. With heavy breathing (hyperpnea), which might happen during strenuous exercise, the stress response, or an asthma attack, the body uses accessory muscles to speed up inhalation and exhalation.

During an active exhalation, the abdominal muscles contract strongly, causing the anterior and lateral rib cage to be pulled downward. The contraction of the abdominal muscles decreases the size of the rib cage and increases abdominal pressure, and the diaphragm is pushed back into its fully domed shape. The volume of air at the end of a forced exhalation will be less than the resting functional residual capacity. However, the lungs cannot be completely emptied. The amount left after maximal exhalation is known as the *residual volume*, which functions to keep the alveoli open. A sneeze is an example of an involuntary forced exhalation while the yogic breathing exercise kapalabhati is a voluntary one.

As the exhalation becomes active, so, too, might the inhalation. During forced inhalation, inspiratory accessory muscles in the neck and outside of the ribs switch on to speed up the inhalation. The diaphragm will still operate but will be assisted by the accessory muscles, which will stiffen the ribs to allow the diaphragm to pump more quickly instead of building intra-abdominal pressure as it usually does (Aliverti 2016). These accessory muscles accentuate the bucket handle movements, bringing about a greater change in thoracic cavity pressure. As a bucket handle rises, it moves away from the center of the bucket. So it is with our ribs and our spine. The swinging out and up of the ribs creates more space, thus aiding inhale.

Interestingly, interlacing your fingers behind your head allows these accessory muscles to work more efficiently. Think of the image of someone who has just completed a 400-meter race. If they are not doubled forward, they probably have their hands behind their head to help with forced inspiration.

While the accessory muscles are useful for increased ventilation, habitually using the accessory muscles as part of the primary muscles of breathing is considered dysfunctional and can create problems, which is discussed later in regard to dysfunctional breathing.

ATMOSPHERIC AIR AND EXPELLED AIR

Knowing that we inhale oxygen and exhale carbon dioxide, one might wonder how cardiopulmonary resuscitation, where you exhale into the mouth of a person who is not breathing, works. The answer lies in the composition of exhaled air.

Atmospheric air contains, by volume, approximately 78 percent nitrogen, 21 percent oxygen, 0.9 percent argon, 0.04 percent carbon dioxide, and small amounts of other gases as well as variable amounts of water vapor. In contrast, exhaled air contains, by volume, 4 to 5 percent carbon dioxide, which is about a hundredfold increase over the inhaled amount. The volume of oxygen is reduced by a small amount, 4 to 5 percent, compared to the oxygen inhaled, thus leaving a substantial amount of oxygen in each exhale. The typical composition of exhaled air is 78 percent nitrogen, 13 to 16 percent oxygen, and 4 to 5.3 percent carbon dioxide, as well as 5 to 6.3 percent water vapor and other trace gases.

In a room with poor ventilation, carbon dioxide levels will slowly rise, making the air feel stuffy and making the people inside feel sleepy, lethargic, or less focused. A 2016 study looked at the effect carbon dioxide has on sleeping patterns and cognitive performance. In a university dormitory setting, the researchers opened the window in some rooms to vent carbon dioxide while other subjects slept with windows closed. The researchers recorded the carbon dioxide levels. They found that the students who slept with the window open during the night, which lowered carbon dioxide levels in the room, slept significantly better than those with elevated levels. They also performed better on cognitive tests the day after and reported that the air felt fresher (Strøm-Tejsen et al. 2016).

Kapalabhati Stops the Aging Process

Kapalabhati is a breathing exercise where the practitioner makes successive, rapid, forceful exhalations through the nose or, less commonly, the mouth. Every Sivananda class starts with it; most kundalini classes feature it; some Jivamukti and Rocket Yoga classes suggest it during Fish Pose (Matsyasana) or at other times; and most hot yoga classes finish with it. With *kapala* meaning skull in Sanskrit and *bhati* meaning light (as in perception and knowledge), kapalabhati is often translated as skull-shining breath. It is performed by exhaling forcefully and then inhaling passively. Hence, the inhalation is slightly longer than the exhalation and an inhale–exhale cycle might last one second or even less.

Articles and videos on the Internet often have titles like, "Release Toxins With Kapalabhati Breath." A blog from Gaia suggests that kapalabhati "fills your stuffy skull with fresh air" (Paschall 2013, para. 3). A lay media site says that "kapalabhati... can detoxify your body almost entirely" and can "protect your lungs against toxic air" (Femina 2020, para. 12; para. 13). The website Yoga International claims that kapalabhati "renews body tissues and helps to arrest old age" (Sovik n.d., para. 5). Regularly appearing on Indian television, Baba Ramdev, one of India's most famous gurus, has claimed that his yoga practices, which feature kapalabhati as a major element, can treat the "curable disease" of homosexuality (Wilson 2009, para. 2; Pradhan 2015).

While some of the aforementioned claims lack sense, we can apply a few physiological principles to estimate the benefits of kapalabhati breathing. As this exercise requires a rapid and sustained effort by the abdominal muscles, the abdominals might experience increased tone. Similarly, muscular contractions release heat, so it stands that performing kapalabhati might generate thermogenesis (heat production). Additionally, kapalabhati demands a sustained focus of the mind, thus bringing the mind's awareness to the breath. Finally, as it is a form of controlled hyperventilation (which is explored later in this chapter), kapalabhati likely makes longer breath holds possible immediately following the exercise as carbon dioxide is off-loaded.

Practicing kapalabhati likely has other benefits, such as creating a feeling of invigoration, but it is not—and cannot be—a panacea for all ailments. As for kapalabhati stopping the aging process, it is probably as effective as snake oil.

GOOD, WHOLESOME OXYGEN?

Considering the 2016 study by Strøm-Tejsen and colleagues and the general physiology of breathing, it is easy to think that we cannot get enough oxygen and that carbon dioxide, which is a waste product of cellular respiration, serves no purpose. Indeed, many yoga teachers talk about oxygen as if it is a wholesome, nutritive element with cues such as "Headstand helps oxygenate the brain" (see the sidebar "Myth or Fact? Yoga Inversions Bring More Blood to the Brain and Stimulate the Pineal Gland" on page 45). Think of oxygen bars, which were a short-lived fad in the 1990s, or the practice of providing medical oxygen to patients who are ill. In fact, oxygen is a highly

And Don't Forget to Breathe!

Consider how the diaphragm is the roof of the abdominal cylinder and how it presses down onto the abdomen as the side walls and pelvic floor engage to also help stabilize the torso and spine. It is little wonder that yogis occasionally hold their breath during a challenging asana and the teacher shouts, "Don't forget to breathe!" But is holding the breath during yoga the cardinal sin it is painted out to be?

Powerlifters use a technique known as the *Valsalva maneuver,* where they inhale and brace their core, then try to exhale but stop the air with their vocal folds, which regulate airflow through the throat. Sprinters also hardly breathe during a short race. In a 100-meter race, an Olympic sprinter will often hold their breath the whole time (which is only 10 seconds) or just take small breaths of air. A breath hold can be a useful technique for a mountain climber when steady IAP is essential for a difficult move. In these examples and many others in the world of physical activity, a breath hold is an important aspect of building IAP. If those athletes were to follow that classic yoga teacher cue, "Don't forget to breathe!" not only could their performance be negatively affected, but in the case of lifting, it could possibly be dangerous.

But yoga is not, of course, an Olympic sport—at least, not yet! Nonetheless, consider Chair Pose (Utkatasana). Because of the torso's angle of inclination, exacerbated by lengthening the arms, which creates a very long lever from the center of mass, Chair Pose is by many standards a challenging pose, especially when held for 5 to 10 breaths. In fact, Chair Pose is very similar, biomechanically, to the sitting phase of an Olympic snatch (figure 3.6). The Olympic lifter will probably use a breath hold, though, while the yogi will probably be cautioned to never hold their breath.

Perhaps, for a beginner who almost never lifts their arms overhead and has very poor tone in their spinal extensors, holding the breath in Utkatasana is allowing them to maintain the pose for what the teacher considers five lengths of breath—and we can probably all think of a teacher whose count to five seems inordinately long.

a   b

FIGURE 3.6 Though the yogi is not carrying any load in her hands in *(a)* Chair Pose, the position is biomechanically similar to *(b)* the sitting phase of an Olympic snatch.

Do you ever struggle to breathe in a backbend? Consider the diaphragm's connection with core musculature in a backbend like Low Lunge (Añjaneyāsana; figure 3.7). Some of the muscles being stretched include the iliopsoas, the core musculature including the rectus abdominis and transversus abdominis as well as the intercostal muscles. In a resting position, such as lying down, these structures would be relaxed and able to move freely with each breath. However, in a pose like Añjaneyāsana, the iliopsoas, core musculature, and intercostal muscles are under tension and not able to move as easily. Even the rib cage is restricted in its ability to move freely. Someone with a very mobile spine might be able to do backbends with a lot of ease and not much change to their breathing. For someone who is tighter, their whole muscular system will be tighter, affecting the breath even more. To say that breathing should be easy in each pose would not be fair. Rather, to expect that each pose will affect the breath in a different way would be fair. For a student, hearing that it is okay if you find it challenging to breathe in a pose is much less alienating than being told that they should be able to breathe easily throughout.

While there is probably nothing wrong with encouraging your students to breathe through a pose, perhaps some compassion would be well advised instead of a shouted command of "Don't forget to breathe!" Perhaps a better cue would be to offer this as a question rather than a commandment: "Can you continue to breathe in this pose? Can you use ujjayi to help steady your abdomen instead of a breath hold?"

Perhaps there is even a place for sometimes holding the breath in a yoga practice. As Handstand (Adho Mukha Vrksasana) is a pose that requires a very steady intra-abdominal pressure, maybe holding the breath as you lift into it might help. You can always then breathe again once you are proudly balancing. Similarly, some people, especially those who are tighter, might benefit from a breath hold while coming up into Wheel Pose (Urdhva Dhanurasana), then resuming breathing once up there.

So, should we never hold our breath in yoga? It seems, as is often the case with the human body, the best answer is probably *it depends*.

FIGURE 3.7   Structures such as the abdominal and intercostal muscles are tensioned in Low Lunge (Añjaneyāsana), which, for someone with tighter muscles, might strongly affect their breathing.

corrosive element that reacts with most substances it contacts, creating oxides. The rust that appears on metal left outside is the result of oxidation, the process of oxygen reacting with the metal. The reactivity of oxygen is a crucial part of the instantaneous filling of airbags in cars to create oxides. Fire is the rapid oxidation of a material in a process known as *combustion*. While oxygen is a necessary component for life as we know it, having too much oxygen in the body can be dangerous, creating oxidative stress, one of the reasons we need to eat antioxidants in our diet! Even though our oxygen needs increase with increased effort, breathing is more about the balance of gases in the body than it is about taking in as much oxygen as possible. As we will see later in this chapter, imbalance can be brought about through hyperventilation.

Breathing and Back Pain

Considering the diaphragm's role in IAP and hence spinal stabilization, it is not so surprising that there may be a link between low back pain and compromised breathing. Vostatek and colleagues (2013) at the Czech Technical University in Prague used magnetic resonance imaging to look at the movement of the diaphragm in 17 people with chronic low back pain with identified spinal structure disorders from disc protrusion to spinal arthritis (spondylosis) in comparison to 16 of their healthy counterparts. They found that when a load was applied to the lower limbs, the subjects with the spine disorders were significantly less able to maintain respiratory diaphragm function, meaning they were significantly less able to breathe as easily or efficiently as the healthy subjects, while the healthy subjects showed more stable parameters of their breathing and posture. In other words, this small study showed a link between diaphragm cooperation and low back pain, yet another reason someone might struggle to breathe in a challenging yoga pose.

The Truth About Belly Breathing

Yoga teachers often identify belly breathing as the holiest type of breath. Belly breathing occurs when the abdomen is displaced outward as a result of the diaphragm pressing down onto the abdominal contents. (Just to be clear: Even though asking someone to breathe down into the belly can be a useful and valid cue, no air is of course descending into the belly, only into the lungs.) Some incorrectly think that belly breathing is synonymous with diaphragmatic breathing, but, in fact, one can breathe diaphragmatically with only the chest moving while the belly remains still.

Author Bernie Clark has made it his mission to share with the yoga community the vastness of individual variation. In *Your Spine, Your Yoga*, Clark (2018) considers the variations of rib cage structure between the sexes and between ages, citing a study by Bellemare, Jeanneret, and Couture (2003), which found that the ribs of the women in the study pointed down significantly more than the men's. Presuming that the subjects were representative of the wider population, women's ribs are more likely to move during respiration than men's ribs. Women's rib cages can achieve a greater volume expansion. Evolutionarily, this may suit the lack of abdominal displacement possible during pregnancy.

The ribs of the elderly and newborns also point outward more than downward, so their ribs cannot expand outward during inhalation, and they both will be inclined to belly breathe. Think of a baby lying on their back and how much the belly distends outward on inhalation. This is not because they are breathing correctly and adults are breathing poorly; rather, our bodies naturally change through the course of our lives. Therefore, seeing someone's chest rise as they breathe does not necessarily mean that they are breathing incorrectly or that they have dysfunctional breathing.

The three-part breath is where the teacher guides the student to first breathe down into the belly, then the middle of the chest, then up into the collar bones. Some have likened it to filling a glass with water. However, when it comes to our lungs, air does not follow the principles of fluid dynamics. It follows the principles of aerodynamics. Imagine blowing up a balloon and trying to fill it from the base of the balloon and moving upward. Of course, that is impossible.

In a person with a complete upper spinal cord injury (SCI), the abdominal muscles are completely paralyzed (Estenne, Pinet, and De Troyer 2000). As a result, the belly distends with each inhale, which makes the rib cage contract, which, as discussed earlier, is paradoxical breathing. The SCI patient's breathing therefore becomes inefficient, and it can be difficult for them to increase breathing rate or deepen their breath, as is required when performing exercise. Though the belly breath is often regarded as the best breathing technique in yoga, belly breathing is more complicated and might not be best suited for a dynamic yoga practice.

UJJAYI BREATHING

A breathing technique used in many styles of yoga is ujjayi pranayama, which means victorious breath and, in the West, is sometimes called *ocean breathing*. While some teachers might describe ujjayi in very esoteric terms, biomechanically it is the same thing as whispering.

Anatomy of Ujjayi

The Adam's apple—the bump in the front of the neck that is more prominent in men— is formed of the thyroid cartilage, which is connected to the larynx, or voice box. Within the voice box are the vocal folds, commonly referred to as the *vocal cords* (figure 3.8). During relaxed breathing, the vocal folds will be opened, or abducted. The space between the vocal folds, where the air passes, is called the *glottis*. During voiced phonation, which is the process of producing sound for speaking or singing, the vocal folds touch together (they are adducted) and vibrate, thus creating sound as the air rushes past them. With whispering, the vocal folds open to just about 25 percent of maximal opening, usually forming a small triangle at the posterior aspect (Ball and Rahilly 2003). During whispering, airflow is strongly turbulent, which creates the hushing sound we know.

Partially constricting the airway with ujjayi breathing (whispering) helps to create a steadier degree of IAP. Additionally, air friction creates heat so the larynx might heat up slightly. Finally, if ujjayi is maintained on the inhalation, which is certainly possible, the diaphragm will have to work harder to contract and draw air in while the expiratory muscles—that is, the deep abdominal muscles—might have to engage during the exhalation. All these factors will certainly help to, as many teachers say, create heat on the inside.

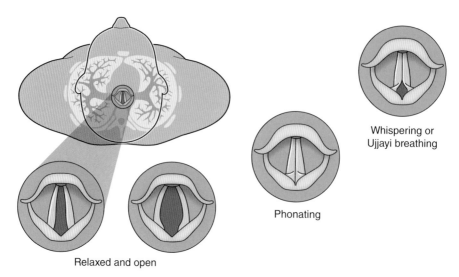

Whispering or
Ujjayi breathing

Phonating

Relaxed and open

FIGURE 3.8 The vocal folds and the glottis while relaxed and open, phonating (speaking), and whispering, or ujjayi breathing.

Ujjayi breathing is not exclusive to yogis. People use ujjayi breathing—or airway constriction with the means of increasing IAP—without ever hearing its name or attending a yoga class. During core work like double leg raises, when you want your intra-abdominal pressure steady, even nonyogis frequently switch on ujjayi breathing. The same can be true when someone is holding a heavy object.

In traditional ashtanga yoga, the practitioner is instructed to maintain ujjayi breathing from the beginning of the practice until the start of Savasana. As Kaminoff and Matthews point out in *Yoga Anatomy*, however, ujjayi breathing might not suit every yoga asana or every yoga style. They say, "Because the ultimate goal of yoga breath training is to free up the system from habitual, dysfunctional restrictions, the first thing we need to do is free ourselves from the idea that there is a single correct way to breathe" (Kaminoff and Matthews 2007, p. 20). Saying that yoga helps with dysfunctional restrictions, though, implies that some, or even many, of us have dysfunctional patterns.

Benefits of Ujjayi

The Internet is rife with pseudoscientific-sounding benefits to ujjayi breathing. At the time of writing this chapter, the Wikipedia entry on ujjayi breathing, which lacked citations, claimed the following: "This breathing technique enables the practitioner to maintain a rhythm to his or her practice, take in enough oxygen, and helps [sic] build energy to maintain the practice, while clearing toxins out of the bodily system" (Wikipedia 2020).

While it makes sense that ujjayi can help maintain a certain rhythm to one's practice, the claims that it will also ensure you have enough oxygen and clear the "bodily system" do not have much of a scientific basis. As for blood oxygenation, ujjayi might reduce the amount of oxygen in our blood because of the slower inhalation but that is not a bad thing. As for detoxifying the body, detoxification is a cellular process that happens with the liver and lymphatic system, and, while evidence exists that shows exercise in general helps make our liver more efficient, there is no evidence or logic to ujjayi breathing significantly doing so. (For further information about detoxification, see chapter 8 on the digestive system.)

Other logical benefits to ujjayi breathing include slowing down the breath, which can have a calming effect on the body; maintaining awareness of the breath as ujjayi requires a continuous conscious effort; and helping to create a steady degree of IAP, which stabilizes the spine and midsection, useful when practicing challenging asanas.

The right way to breathe depends on the task being performed, whether that is sprinting, relaxing, or practicing the primary series of ashtanga. So ujjayi breathing would not suit all types of yoga or all asanas within each practice.

NOSE BREATHING AND MOUTH BREATHING

Recently, attention has been paid to mouth breathing versus nose breathing. In his book *Breath*, James Nestor presents a significant amount of research showing how nose breathing is far superior to mouth breathing (2020).

The benefits of nose breathing have been explored and described in the scientific literature since at least the 1950s (Cottle 1958). The nose has at least 30 health-protecting functions, and nose breathing, as opposed to mouth breathing, is understood by McKeown, O'Connor-Reina, and Plaza (2021) to

- warm and humidify inhaled air;
- filter air, reducing exposure to foreign substances and pathogens;
- increase oxygen uptake and circulation;
- slow down breathing rate;
- improve lung volume;
- support the correct formation of the teeth and mouth; and
- lower your risk of snoring and sleep apnea (a disordered, stop–start breathing condition).

Nose breathing also imposes approximately 50 percent more resistance to the air stream than mouth breathing. While air resistance might not sound like a good thing for breathing, such resistance results in 10 to 20 percent more oxygen uptake because of the biomechanics of slowed breath. Additionally, sufficient nasal resistance during inhalation allows the diaphragm to work more efficiently and maintains elasticity of the lungs. Furthermore, this resistance to breathing slows the breathing rate, allowing our bodies to retain an ideal amount of carbon dioxide, which is not just a waste product but serves several important roles in the body.

Nasal breathing also allows for the mixing of inhaled air with a gas called *nitric oxide*, which is produced in the nasal sinuses. Essential for overall health, nitric oxide improves the lungs' ability to absorb oxygen; increases our ability to transport oxygen throughout the body; allows blood vessels to dilate through the relaxation of vascular smooth muscle; and serves antifungal, antiviral, antiparasitic, and antibacterial purposes.

Meanwhile, mouth breathing is associated with a number of problems, including increased allergic reactions to allergens, asthma, bad breath, tooth decay, gum inflammation (gingivitis), snoring, and teeth or jaw abnormalities. Truly, the benefits of nose breathing seem manifold, and breathing through the nose during a yoga practice is a good idea. However, as is often the case with physiology, there are exceptions to this rule. Because nose breathing is slower, it might not be sufficient for high-intensity exercise when oxygen needs are much greater than at rest. In such cases, it is best to let your body decide how it needs to breathe. For chronic mouth breathers, though, the slow and mindful breathing practices of yoga can be very powerful.

THE PSYCHOLOGICAL AND PHYSIOLOGICAL EFFECTS OF PRANAYAMA AND SLOW BREATHING

At the heart of most pranayama exercises is the act of slow breathing, and thankfully, there has been a decent amount of research on slow breathing. Between 12 and 20 breaths per minute is considered the normal respiratory rate for healthy adults (Flenady, Dwyer, and Applegarth 2017). So, anything less would be considered a consciously slowed-down breathing rate.

While researchers do not yet understand the exact reasons why, it is well established that the experience of emotions has whole-body effects (Cacioppo et al. 2000) and that breath is strongly linked with emotions in part via the autonomic system (Kreibig 2010). We can all attest to this. Feeling low or anxious elicits a certain breathing response in the body, while feeling elation elicits another. It is little wonder that the original Latin word for inspiration has taken on an emotionally evocative meaning while the original Latin word for expiration has taken on quite another.

The autonomic nervous system is the part of the nervous system that supplies the internal organs such as the blood vessels, stomach, intestine, liver, kidneys, bladder, heart, and digestive glands, thus influencing many involuntary functions such as heart rate, blood pressure, respiratory rate, digestion, and sexual arousal. It has two primary branches: the sympathetic, commonly referred to as the *fight-or-flight response* but also necessary for exercise and many everyday activities including sweating; and the parasympathetic, commonly called the *rest-and-digest response*, which elicits feelings of relaxation. (For more detailed information about the autonomic nervous system, see chapter 2 on the nervous system.)

Several reviews have looked at the available evidence on slow breathing to better understand its psychological and physiological effects. The reviews generally agree that slow-breathing practices like pranayama activate a parasympathetic, or calming, effect in the practitioner. One study in particular was a 2018 systematic review of all the literature on slow breathing. In this review, Andrea Zaccaro at the University of Pisa and colleagues looked at the effects of slow breathing (10 breaths per minute or fewer) in healthy people and the possible psycho-physiological mechanisms behind

Bad Posture Leads to Poor Breathing

Is there any truth to your mother telling you to sit up straight so you can breathe properly? Different yoga postures can greatly affect our breathing (see page 68 of this chapter), so it would make sense that our everyday posture would also affect our breathing. So, the real question is: Does bad posture lead to poor breathing?

Tight neck and chest muscles, such as the intercostal and pectoralis muscles, can restrict the free movement of the rib cage. Additionally, if you are slouching forward, your abdominal organs will be more compacted and not able to move as freely as the diaphragm descends. For this same reason, we can sometimes find it difficult to breathe deeply after a big meal; the abdominal contents cannot be as easily displaced as before the meal. The same is the case for a pregnant woman where the abdomen cannot be moved very much because of the immovable bundle of joy within.

While we wish to emphasize the point that there is not one universal proper posture, some research supports the idea that posture affects breathing. Szczygieł and colleagues (2015) found that the farther forward the head is during breathing, the less the lower ribs move. This is not a surprising finding really. It just suggests that our posture affects our breath.

A small number of studies, including one from South Korea (Jung et al. 2016), have connected use of electronics including smartphones and laptops with forward head posture (called *text neck*) and found a connection between forward head posture and reduced lung function.

Dimitriadis and colleagues (2014) found that people with chronic neck pain do not have optimal lung function. The authors found a correlation between compromised lung function, weak neck muscles, and kinesiophobia, or fear of movement. They suggest that both pain and kinesiophobia may alter neck biomechanics, further contributing to the development of respiratory dysfunction.

So, the literature might support the idea that our sitting and standing posture, including our neck position, can affect breathing but probably not to the extent that many blogs might make you believe. Also, this does not mean that slumping onto the sofa is inherently bad; sometimes nothing can be better after a long day. But every now and again, it is certainly a good idea to check in with how you are breathing and how you are feeling. After all, what does yoga teach us if not awareness of breath and self?

those effects. In an article titled *How Breath-Control Can Change Your Life* (2018), Zaccaro and colleagues cited evidence that slow-breathing techniques increase mental and physiological flexibility, parasympathetic activity, and central nervous system activities related to emotional control and psychological well-being.

The researchers found that slow-breathing techniques at the rate of six breaths per minute had the most reliable associations with parasympathetic activity, as measured with increased heart rate variability; increased alpha waves, as measured with electroencephalogram; and positive psychological and behavioral effects (Zaccaro et

---

TRY IT YOURSELF: Breathing Rate

Grab a stopwatch (most smartphones have a stopwatch app) and, without trying to change your breathing, measure how many breaths you are taking per minute. Take note of how you feel.

Next, consciously slow down your breathing, aiming for as little as three or four breaths per minute. You could try this as a seven-second inhalation, a very brief retention, and a seven-second or longer exhalation. Note how you feel when you breathe at that rate.

---

al. 2018). Slow breathing has also been shown to cause near-complete inhibition of sympathetic activity (Seals, Suwarno, and Dempsey 1990), meaning that the neurons connected to the activation mode (fight-or-flight mode) are almost completely switched off.

Zaccaro and colleagues (2018) included studies that tested yogic pranayama, a Zen breathing practice, and general slow-paced breathing techniques. Unfortunately, because many of the yoga-based studies had low-quality methodologies, only one pranayama study could be included. The researchers found, as previous researchers have, that the "brand" of breathing technique did not seem to matter much because, "In our opinion, it is possible that certain meditative practices and slow-breathing techniques share, up to a point, similar mechanisms" (Zaccaro et al. 2018, p. 12). Conversely, irregular breathing such as shallow or deep, rapid breathing with periods of apnea, which is the cessation of breathing, leads to increases in sympathetic activity (Leung et al. 2006).

All these findings offer prime examples of how breathing directly affects our autonomic nervous system. As for the mechanism behind this connection, Zaccaro and colleagues (2018) suggested that the epithelium (the skin) on the inside of the nostrils may play an important role in the act of slow breathing causing autonomic changes. Evidence from both animal models and humans support the hypothesis that nostril-based breathing, which stimulates certain receptors in the nostril epithelium, could be one of the pivotal mechanisms behind our breath affecting our mood.

While many people tend to feel more relaxed when taking slower breaths, this might not be so for everyone. Similarly, telling someone to "just breathe" or to "just relax" might not be as comforting as the speaker might intend. It is important to learn what works for us and what does not so that we can have these tools to calm ourselves down in difficult situations. In that way, understanding our breathing helps us understand ourselves.

CONDITIONS OF THE RESPIRATORY SYSTEM

As breathing is such an important aspect of yoga, some basic information on some common respiratory conditions can be useful. Yoga is often described as a healing practice. The conditions dysfunctional breathing, hyperventilation syndrome, asthma, and chronic obstructive pulmonary disorder are explored in the following sections, and we discuss whether evidence exists to support yoga as a treatment for these maladies.

Dysfunctional Breathing

It is a very common idea among yoga teachers and practitioners alike that most people's breathing is suboptimal. Additionally, yoga teachers often advertise their breathing workshops with messages like "Learn how to breathe properly" or rather authoritatively "You have been breathing incorrectly your whole life," a message that might scare almost anyone into handing over their money. Dysfunctional breathing is a recognized medical condition, but are we all suffering from it?

Dysfunctional breathing (DB) is defined as an alteration in the normal biomechanical patterns of breathing that result in intermittent or chronic symptoms. DB is recognized by the asthma guidelines of the British Thoracic Society and the Scottish Intercollegiate Guidelines Network, as well as the Global Initiative for Asthma. The condition has also, problematically, been described as *hyperventilation syndrome*, *disproportionate breathlessness*, *behavioral breathlessness*, *anxiety-related breathlessness*, *psychogenic functional breathing disorder*, and *somatoform* (as in psychosomatic) *respiratory disorder*—all describing essentially the same problem (Barker and Everard 2015). Many of these terms have a significant psychological component, and this perceived link between psychological dysfunction and DB probably causes many physicians to avoid diagnosing the condition (Barker and Everard). Being poorly understood, dysfunctional breathing is often underdiagnosed or misdiagnosed in clinical practices worldwide, which can deprive a person of adequate treatment (Vidotto et al. 2019).

DB is most common among individuals with asthma, and a central component of it is pattern-disordered breathing, in which a person uses the upper chest wall and accessory muscles as the primary respiratory pump instead of using the diaphragm as with normal relaxed breathing. This is the same type of breathing so many yoga teachers seem to be concerned with and lead workshops about. But when someone has DB, they know.

Someone with DB will experience one of the following: sporadic bouts of hyperventilation; periodic deep sighing; thoracic dominant breathing; forced abdominal expiration; or thoracoabdominal asynchrony, meaning there is a delay between rib cage and abdominal contraction, resulting in ineffective breathing mechanics. With those breathing patterns, a person will experience other symptoms such as increased breathing rate, breathlessness, lightheadedness, dizziness, tingling sensations, muscle spasms, or numbness.

While yoga teachers might be trying to be helpful, they should probably stop telling their students that they do not know how to breathe. While it is entirely possible that someone might enter a yoga studio with DB, it is not the place of the yoga teacher to diagnose it. Furthermore, using fear-based language to suggest that everyone is breathing incorrectly can contribute to someone's anxiety instead of helping to relieve it—the opposite of the goal of someone whose aim is to help others.

Though yoga teachers do not have the authority to diagnose, they still have an important role to play in guiding their students' breathing. In fact, when done well, this is probably the best gift we can impart to someone else; few other movement modalities focus on breath awareness to the extent that yoga does. Indeed, simply bringing someone's awareness to their breath is a powerful tool. The simplest meditation we can offer our students is to simply relax and observe their breathing free of judgment about the quality of their breathing and free of language that they are not using their diaphragm correctly. (A sample breath meditation, "Slow, Relaxed Breath of Gratitude," is provided at the end of this chapter.) While a controlled breath practice or slowed breath can be beneficial, perhaps sometimes all we need to do is relax and get out of the way of the breath.

We are still a long way from fully understanding DB. The classifications, definitions, and diagnostic criteria for DB were still being discussed in a review as recently as 2019 (Vidotto et al.). Considering that medically trained professionals are misdiagnosing and underdiagnosing dysfunctional breathing, clearly it is not the role of the yoga teacher to diagnose people as dysfunctional breathers or to tell them they have been breathing incorrectly their whole life. Yoga teachers are, however, well placed in guiding their students through pranayama, slow-breathing practices, and general breath awareness, all of which can have profound benefits to well-being.

Overbreathing and Hyperventilation Syndrome

Overbreathing, or hyperventilation, is characterized by very rapid breathing, which usually disturbs the balance of carbon dioxide and oxygen. It is the breathing pattern seen by someone having a panic attack and has strong psychological links, both in its causes and its effects. In this type of breathing, the inhalation is longer than the exhalation, and a person who is overbreathing usually looks like they are gasping for breath. While the individual might feel that they cannot get enough oxygen, the problem is actually that they are not retaining enough carbon dioxide. Within our blood vessels, we have many receptors telling our brain how much carbon dioxide is present. Carbon dioxide is a waste product of respiration but is also necessary to help oxygen unbind from hemoglobin so that the oxygen can reach various organs. (For more information about hemoglobin and the transport of oxygen, see chapter 4 on the cardiovascular system.) Carbon dioxide also affects the pH or acid–base balance of our blood. In hyperventilation, the body eliminates more carbon dioxide than it can produce; this is called *hypocapnia* (*hypo* means reduced, and *capnia* refers to carbon dioxide in the blood). This decrease in carbon dioxide leads to an increase in blood pH, meaning that the blood becomes more alkaline, and oxygen cannot unbind from hemoglobin, which means decreased oxygen to the tissues. The symptoms of respiratory alkalosis include dizziness, tingling in the lips, hands, or feet, headache, weakness, fainting, and seizures. In extreme cases, it may cause carpopedal spasms, a flapping and contraction of the hands and feet. Hence, a person can feel lightheaded and dizzy because the tissues of the brain cannot receive adequate oxygen (Gardner 1996).

While hyperventilation can be executed voluntarily, as is the case with kapalabhati pranayama, hyperventilation syndrome describes the respiratory disorder of a hyperventilation episode occurring involuntarily. A strong correlation between panic disorder and hyperventilation syndrome exists and is well established in the scientific literature (Sikter et al. 2007). Contemplative practices, such as yoga, meditation, and tai chi, which focus on slow breathing and long exhalations, are known to induce respiratory vagus nerve stimulation, or kick-start the calming rest-and-digest influence of the parasympathetic nervous system (Gerritsen and Band 2018). Therefore, though there is a lack of studies showing the effectiveness of breathing exercises for dysfunctional breathing and hyperventilation syndrome in adults (Jones et al. 2013)

Hyperventilation as a Tool

Free divers—people who deep dive with no scuba equipment—often perform breathing exercises that mimic hyperventilation prior to a dive to make their blood alkaline, oxygen saturated, and carbon dioxide deficient. As the free diver then holds their breath, the opposite occurs: Carbon dioxide levels increase in the blood, which makes the blood more acidic, and oxygen levels deplete. By hyperventilating themselves, free divers put themselves in an alkaline state, which then gives them a buffer before going to neutral and finally acidic.

Wim Hof, also known as "The Iceman," is an extreme athlete who has set Guinness-recognized world records for swimming under ice, prolonged full-body contact with ice, and running a half-marathon barefoot on ice and snow. A central tenet of the Wim Hof method is a breathing practice of controlled hyperventilation followed by a breath hold. According to Hof, this breathing technique helps to calm a person before or during an extreme event like plunging into an ice bath for 60 minutes (Wim Hof Method 2020).

Hyperventilation has also been used as a therapeutic practice, known as *Holotropic Breathwork*, which is intended to bring the practitioner into an altered state of consciousness to help with emotional healing and personal growth (Holmes et al. 1996). Accompanied by music, it involves breathing at a fast rate for minutes to hours to transcend the body and connect with one's true self then creatively reflect on the experience. It is meant to be guided by someone trained in this emotional release modality, and at least three studies have found positive outcomes, including emotional catharsis, internal spiritual exploration, and higher levels of self-awareness as well as temperament changes where practitioners reported less tendency to be needy, domineering, and hostile.

These examples carry some risk and might not be appropriate for everyone, especially someone with breathing issues. What these examples highlight, however, is that something that is usually perceived negatively—hyperventilation—can be a beneficial tool when used in a controlled way.

---

or even children (Barker et al. 2013), it makes physiological sense that yoga might help to reduce episodes of involuntary hyperventilation by increasing parasympathetic activity and one's ability to handle stress (De Couck et al. 2019).

Asthma

Being that it is such a common respiratory disease, we probably all know someone who has asthma, or we have at least seen someone using an inhaler. Asthma is a persistent inflammatory disease in which the airways of the lungs narrow and swell, making breathing difficult. This can trigger coughing, wheezing, chest tightness, and shortness of breath. Its severity can range from being a slight nuisance for some to a debilitating and life-threatening condition for others. An asthma attack may occur a few times per week or even a few times per day, and, depending on the person, symptoms may be worse at night or during exercise.

While it is not fully understood why some people get asthma and others do not, asthma is thought to be caused by both genetic and environmental factors. Various irritants and substances that trigger allergens can trigger signs and symptoms of asthma. Asthma triggers are different from person to person but have been noted to include

- airborne allergens, such as pollen, dust mites, mold spores, pet dander, or particles of cockroach waste;
- respiratory infections such as the common cold;
- physical activity;
- cold air;
- air pollutants such as smoke and the spraying of pesticides;
- certain medication including beta-blockers, aspirin, and nonsteroidal anti-inflammatory drugs; and
- sulfites and preservatives added to some types of foods and beverages including processed meat, shrimp, dried fruit, processed potatoes, beer, and wine; and
- strong emotions and stress.

People with a family history of allergies are more likely to develop asthma. These can include asthma itself, eczema, hay fever, or other allergies. Finally, smoking plays an important role in asthma.

A 2014 review of 14 randomized controlled trials with a total of 824 participants looked at the effect of yoga on symptoms, lung function, and quality of life for people with asthma (Cramer et al.). The authors determined that they could not find evidence that yoga was effective in improving the outcomes they were measuring, and so yoga could not currently be considered a routine intervention for asthmatic patients. They did, however, conclude that it can be considered an additional intervention or an alternative to breathing exercises for people with asthma.

A different review by Yang and colleagues (2016) of 15 randomized controlled trials with a total of 1,048 participants found moderate-quality evidence that yoga probably leads to small improvements in quality of life and symptoms in people with asthma but that the effects of yoga on lung function were inconsistent. Though they found a small amount of evidence that yoga can reduce medication usage, they determined that most of the studies were flawed in various ways.

A study by Santino and colleagues (2020) looked at different breathing techniques including yoga, Buteyko, and the Papworth method. This study differed from the two aforementioned reviews as it included breathing methods outside of yoga and it of course included more recent research as well. The Papworth method focuses on slow, deep belly breathing while the Buteyko method emphasizes nasal breathing, breath holds, and relaxation. Both methods have similarities with yogic breathing techniques. Some of the studies from the yoga classification of this study included pranayama alone, while others included breathwork alongside asana and meditation. The researchers concluded that breathing exercises may have some positive effects on quality of life, hyperventilation symptoms, and lung function. However, due to methodologies of the studies, the quality of evidence for the measured outcomes ranged from moderate to very low.

So, the evidence that yogic breathing—and other breathing methods—helps with asthma is minimal, and more high-quality studies involving large numbers of participants are needed to make firm conclusions about yoga and asthma. Even then, the type of yoga would matter a lot. A gentle hatha class might have vastly different effects compared to a vigorous vinyasa practice.

Beyond measuring lung function, it is important to also consider one of yoga's most beneficial aspects: how yoga helps us de-stress and relax. There is a lot of anecdotal evidence of asthma sufferers finding relief through yoga, and perhaps it is connected to how yoga can relax us, as stress and anxiety can be a major factor contributing to someone's asthma attacks. Yoga might also help by improving posture, which can help the diaphragm work more efficiently, and by opening the chest muscles, which can make breathing easier.

If you know of someone, perhaps yourself, who has asthma and finds relief from yoga and a lessening of symptoms, you do not need a systematic review of randomized controlled trials to tell you that that is the right thing for you to do.

Chronic Obstructive Pulmonary Disease

Chronic obstructive pulmonary disease (COPD) is a chronic inflammatory lung disease in which airflow from the lungs is obstructed. Chronic bronchitis and emphysema are the two most common conditions that contribute to COPD, and they usually occur together. Chronic bronchitis is inflammation of the lining of the bronchial tube, while emphysema is a condition in which the alveoli of the lungs are destroyed as a result of damaging exposure to cigarette smoke and other substances. Symptoms include breathing difficulty, shortness of breath, cough, mucus production, and wheezing. People with COPD are at increased risk of developing heart disease, lung cancer, and a variety of other conditions. Although the symptoms of COPD tend to gradually worsen over time and can affect a person's daily life, treatment can help keep the condition under control.

Long-term cigarette smoking is the cause for most COPD cases, but other irritants like second-hand smoke, pipe smoke, air pollution, and workplace exposure to dust, smoke, or fumes can also cause COPD. The best way to prevent COPD is to never smoke, or to stop now. Of course, for many people, stopping is easier said than done.

In a review of studies looking at mind–body practices as a drug-free treatment for smoking cessation, Carim-Todd, Mitchell, and Oken (2013) concluded that the studies to date support yoga and meditation-based therapies as candidates to assist smoking cessation. However, because of the small number of studies available and associated methodological problems, the researchers determined that more clinical trials with larger sample sizes and carefully monitored interventions are needed to determine for certain whether yoga and meditation are effective treatments. Even with that caveat, this review provides a good basis for using yoga and meditation to help people stop smoking.

For those who have COPD, whether because of smoking or otherwise, the literature supports the idea that yoga might be a crucial approach to managing the disease. Wu and colleagues (2018) looked at all studies of meditative movement to date, including yoga and tai chi, and in their review and meta-analysis, they determined that it might improve many COPD factors, including exercise capacity, shortness of breath, health-related quality of life, and lung function. As most studies do, they closed by stating that more studies are required to substantiate the preliminary findings, but their review shows a lot of hope for improving the lives of people affected by COPD.

TRY IT YOURSELF: Slow, Relaxed Breath of Gratitude

Lie down with your knees bent and the soles of your feet on the floor. Let your inner knees rest against each other so that you are not balancing your legs in the air. Rest one hand on your tummy and one on your chest. Start by taking a few slow, deep breaths as you let all tension melt into the floor. You are in a position where you do not need to use any muscular effort, so let it all release away. Once you feel fully relaxed, notice the movement in your body as you breathe. Observe the gentle rise and fall of your chest and your abdomen. Your breathing does not have to be fixed or corrected. Think of your breathing as being perfect just as it is, so you can simply relax into the soft undulation of the breath. Take a moment to practice gratitude for each breath and your body's ability to exchange gases and sustain your life force—your spirit.

CONCLUSION

Even though its primary function is gas exchange, breathing has many other roles and has the potential to affect all other systems in the body. While the absorption of oxygen and release of carbon dioxide is vital, breathing is more about balancing those two gases than having as much oxygen as possible. The diaphragm is a powerful muscle that acts as the primary motor of breathing and also plays an essential role, along with other deep core muscles, in intra-abdominal pressure and spine stabilization, which are important in yoga and arguably all physical activities. Hence, our breathing can be affected by our movement and posture, and our movement and posture can be affected by our breathing. Holding your breath might sometimes be the most appropriate choice, depending on the task at hand. Breathing is integrally connected to our psychological state via the autonomic nervous system. While dysfunctional breathing is a real thing, it is probably less common than most people think, and the yoga teacher is not qualified to diagnose such. One of the most valuable things a teacher can offer (or we can explore on our own) is to become aware of your breath, relax into your breath, and practice slowing it down.

Chapter 4

CARDIOVASCULAR SYSTEM

The cardiovascular system comprises the heart, blood vessels, and blood. It is an incredibly intricate and efficient system involved in transporting oxygen and nutrients to our cells, removing waste products from the cells, fighting infection, regulating body temperature, and maintaining homeostasis.

Our understanding of the cardiovascular system has evolved immensely since the 1600s. Prior to that, the concept of blood circulating the body eluded investigators (Aird 2011). The ancient Greeks believed in a dual system of veins and arteries, while Galen, one of the most accomplished of all medical researchers during the height of the Roman Empire, proposed that veins contain blood, whereas arteries contain blood imbued with vital spirits. Blood was seen to slowly ebb and flow rather than to circulate. The question of how blood transfers from the right ventricle of the heart to the left ventricle challenged investigators for the next 1,500 years. William Harvey, born in 1578, believed that arteries and veins both served the same function. His theory of blood circulation is widely recognized as the foundation of modern medicine (Lubitz 2004). Controversy surrounding Harvey's circulation model would persist until Malpighi's discovery of capillaries in 1661. He hypothesized that capillaries were the connection between arteries and veins, allowing blood to flow back to the heart in a circulatory way (Pearce 2007).

The heart itself is often mentioned in yoga classes, with heart-opening sequences being a popular choice for many yoga teachers. There is also widespread belief that yoga is simply good for the heart. In this chapter, we will explore the physiology of the cardiovascular system as well as looking in detail at how yoga affects its functioning.

BLOOD

Blood constitutes approximately 8 percent of adult body weight. The average adult female typically has about four to five liters of blood, while the average adult male typically has five to six. A single drop of blood contains millions of red blood cells, white blood cells, and platelets (figure 4.1).

Plasma is a thick, straw-colored fluid that makes up around 55 percent of our blood. It is mostly water (around 92 percent) and carries mineral salts, nutrients, waste products, hormones, enzymes, and antibodies throughout the body.

Red blood cells, known as *erythrocytes*, are estimated to make up around 25 percent of the total cells in the entire body. They are produced in the red bone marrow at the staggering rate of more than 2 million cells per second (Higgins 2015), and their production is controlled by erythropoietin, a hormone produced primarily by the kidneys. Erythrocytes live up to 120 days in circulation. They contain hemoglobin, a red protein that binds to oxygen and carbon dioxide, transporting them to the appropriate place in the body. Mammalian erythrocytes are unique among animal cells in having no nucleus in their mature, functional state. It is suggested that the biconcave shape of the erythrocyte has evolved out of a necessity to maximize nonturbulent blood flow and minimize the scatter of platelets (Yoshizumi et al. 2003).

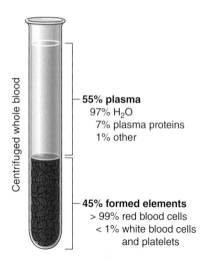

FIGURE 4.1 Constituents of blood.

White blood cells, known as *leukocytes*, are mainly produced in the bone marrow and are involved in protecting us against infection. Leukocytes are far less numerous than erythrocytes and have a much shorter life span. Whereas erythrocytes spend their days circulating within the blood vessels, leukocytes routinely leave the bloodstream to perform their defensive functions in the body's tissues. The most common type of leukocyte is the neutrophil, which is the immediate response cell and accounts for 55 to 70 percent of the total leukocyte count. Neutrophils are remarkably short lived with a circulating half-life of six to eight hours and are therefore produced at a rate of between 50 and 100 billion cells per day (Summers et al. 2010).

Platelets are also made in the bone marrow and are critical to hemostasis, the process of stopping bleeding to keep blood within a damaged blood vessel. Increasing experimental and clinical evidence identifies platelets as important players in other processes including inflammation and tissue regeneration. Platelet-rich plasma derivatives are used in regenerative medicine for the treatment of several clinical conditions including ulcers, burns, muscle repair, bone diseases, and tissue recovery following surgery (Etulain 2018).

HEART

The heart, situated within the rib cage behind and slightly left of the breastbone (sternum) and above the diaphragm, is a muscular pump approximately the size of a fist. The heart of a well-trained athlete can be considerably larger than this.

The muscular tissue of the heart, known as *myocardium*, is made of specialized cardiac muscle. Cardiac muscle shares a few characteristics with both skeletal muscle and smooth muscle (which we will discuss later in the chapter), but it has some distinguishing properties of its own. First, it is specifically designed for endurance. If the average rate of contraction of the heart is 75 contractions per minute, a human heart would contract approximately 108,000 times in 1 day, more than 39 million times in 1 year, and nearly 3 billion times during a 75-year life span. With cardiac muscle there is no possibility of tetany, a condition in which muscle remains involuntarily contracted (think of a muscle cramp in your calf). Finally, cardiac muscle has the property of autorhythmicity—the ability to initiate its own electrical potential to trigger the contractile mechanism.

The heart has four chambers (figure 4.2): the right and left atria and the right and left ventricles. The atria act as receiving chambers while the ventricles serve as the primary pumping chambers of the heart, propelling blood to the lungs or the rest of the body. There are four valves within the heart that prevent the backward flow of blood: The tricuspid valve is located between the right atrium and the right ventricle; the pulmonary valve is located between the right ventricle and the pulmonary artery; the mitral valve is located between the left atrium and the left ventricle; and the aortic valve is located between the left ventricle and the aorta. The flaps of each heart valve are attached to tendinous cords that are sometimes referred to as the *heart strings*.

The first portion of the aorta, the main artery supplying blood to the body from the heart, gives rise to the coronary arteries, which supply blood to the myocardium and other components of the heart. Damaged cardiac muscle cells have extremely limited abilities to repair or replace themselves. In the event of a heart attack (myocardial infarction), dead cells are often replaced by patches of scar tissue. Later in the chapter we will look at how yoga can help to decrease the risk of cardiovascular disease.

The heart has its own conduction system. During contraction of the heart, there is a domino effect: When one fiber receives the signal to contract, the signal immediately spreads to all the muscle fibers of the heart. The contraction is established by the sinoatrial node, a specialized clump of myocardial conducting cells located in the walls

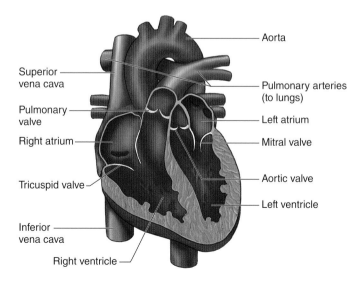

FIGURE 4.2 Chambers and valves of the heart.

of the right atrium. The sinoatrial node is known as the *pacemaker of the heart*. The impulse that is generated from the sinoatrial node then spreads throughout the atria to the atrial myocardial contractile cells and the atrioventricular node, which electrically connects the atria and ventricles.

CIRCULATION

The cardiovascular system has two main divisions: the pulmonary division and the systemic division (figure 4.3). The pulmonary division runs from the right side of the heart to the lungs and back to the left side of the heart. The systemic division runs from the left side of the heart, through the body, and back to the right side.

Cardiac Cycle

The cardiac cycle comprises a complete relaxation and contraction of both the atria and ventricles and lasts less than one second. Beginning with all chambers in relaxation, known as *diastole*, deoxygenated blood flows passively from the superior and inferior vena cava into the right atria, and oxygenated blood flows from the pulmonary veins into the left atria. The atria begin to contract, a process known as *atrial systole*, and actively pump blood into the right and left ventricles. The ventricles then begin to contract, a process known as *ventricular systole*, raising the pressure within the ventricles and pushing open the pulmonary and aortic valves. The deoxygenated blood then moves

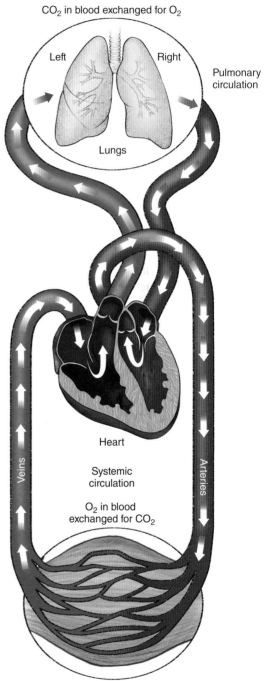

FIGURE 4.3 Pulmonary and systemic circulation.

into the pulmonary trunk from the right ventricle to become oxygenated in the lungs, while the oxygenated blood moves into the aorta from the left ventricle, to begin its journey around the body. The ventricles then begin to relax, which is known as *ventricular diastole*, and pressure within the ventricles drops.

Blood Vessels

The hollow passageway of blood vessels (figure 4.4) through which blood flows is called the *lumen*, meaning opening in Latin. The walls of all blood vessels have three distinct tissue layers, called *tunics*. The innermost tunic is lined with a specific type of tissue called *endothelium*. The substantial middle tunic consists of layers of smooth muscle, a specific type of involuntary muscle, while the outermost layer contains collagenous and elastic fibers.

The arteries closest to the heart have the thickest walls and are known as *elastic arteries* because of their high percentage of elastic fibers, allowing them to distend as needed. Farther from the heart, where the surge of blood has diminished, the percentage of elastic fibers in an artery's wall decreases and the amount of smooth muscle increases. The artery at this point is described as a muscular artery.

There are minute nerves, known as *nervi vasorum*, within the walls of the blood vessels that control the contraction and relaxation of smooth muscle. As smooth muscle contracts, it causes the lumen of the blood vessel to narrow; this is termed *vasoconstriction*. As smooth muscle relaxes, it causes the lumen of the blood vessel to widen; this is termed *vasodilation*.

There is a gradual transition as the vascular tree repeatedly branches. Muscular arteries branch to distribute blood to the vast network of arterioles, which eventually lead to capillaries. A capillary is a microscopic channel, sometimes so small that there is only a single cell layer that wraps around to contact itself. At the capillary level, oxygen and nutrients diffuse from the blood into the surrounding cells and their tissue fluid, known as *interstitial fluid*, in a process called *perfusion*. Waste products that have been produced by the surrounding tissues then enter the capillary system.

Next, deoxygenated blood passes through a series of venules, which gradually become larger veins, and eventually returns to the heart. The flow of blood back to the heart is known as *venous return*. Many veins have one-way valves that prevent the backflow of blood. These valves are present most commonly in veins situated in the limbs or below the level of the heart. In addition to their primary function of returning blood to the heart, veins may be considered blood reservoirs, since systemic veins

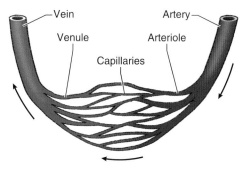

FIGURE 4.4 Blood vessels.

contain approximately two-thirds of the entire blood volume at any given time. When blood flow needs to be redistributed to other portions of the body, the smooth muscle in the walls of the veins constricts. The constriction of smooth muscle in veins is specifically known as *venoconstriction*.

ROLE OF THE CARDIOVASCULAR SYSTEM IN MAINTAINING HOMEOSTASIS

Homeostasis is a term coined by the Harvard physiologist Walter Cannon in 1932. It refers to the tendency of the systems of the body to move toward stability. *Allostasis,* a term coined by McEwen and Stellar (1993), is a dynamic process whereby adaptations occur in the biochemistry and physiology of our internal world to allow us to function optimally in our environment. Allostasis involves changes in the immune, endocrine, and nervous systems in response to short-term or long-term stress. These terms can relate to any biological or physiological function, such as blood pressure, heart rate, core temperature, blood glucose levels, the concentration of electrolytes, and the like.

If you are practicing hot yoga, your rising core temperature will trigger several homeostatic mechanisms, including vasodilation of the blood vessels that are traveling to the periphery of your body, which increases the amount of blood reaching your skin. As blood passes through the vessels of the skin, its heat is absorbed by sweat and dissipated into the environment as it evaporates. The blood returning to the center of your body will then be cooler. In contrast, on a cold day, blood is diverted away from the skin to maintain a warmer body core. Our bodies become acclimatized to heat by increasing the efficiency at which blood is shunted to the skin for cooling. These adaptations occur within a week of being in a hot environment and last several weeks thereafter (Racinais et al. 2015).

Blood also helps to maintain the chemical balance of the body. Proteins and other compounds in blood act as buffers, which thereby help to regulate the pH of the body tissues. Blood plays an additional role in helping to regulate the water content of body cells.

Intrinsic Autoregulatory Processes

There are two basic theories for the regulation of local blood flow when either the rate of tissue metabolism changes or the availability of oxygen changes: the vasodilator theory and the oxygen demand theory (Hall 2015).

There is considerable evidence that a greater rate of metabolism or a reduced availability of oxygen or other nutrients to actively metabolizing cells surrounding arterioles causes them to release vasoactive substances that cause vasodilation of the blood vessels. This is termed the *vasodilator theory*. These metabolic mechanisms ensure that the tissue is adequately supplied with oxygen and that products of metabolism such as carbon dioxide are removed.

Although the vasodilator theory is widely accepted, many physiologists favor an alternative—the *oxygen demand theory*. In the absence of adequate oxygen, it is

Can Yoga Improve Your Circulation?

Blood is easily pumped to our lower limbs from the heart via large arteries; however, returning blood to the heart is not such an easy process. The walls of the veins are considerably thinner, and their lumens are correspondingly larger in diameter compared to arteries, allowing more blood to flow with less vessel resistance. But by the time blood has passed through capillaries and entered venules and then veins, the pressure initially exerted on it by heart contractions has diminished significantly. Venoconstriction is also much less dramatic than the vasoconstriction seen in arteries and arterioles and may be likened more to a stiffening effect of the vessel wall, rather than significant constriction per se. The venous system is also normally working against gravity to return blood from the lower limbs to the heart.

Yoga can help to improve venous return in three main ways: (1) by improving the efficiency of the skeletal muscle pump; (2) by improving the efficiency of the respiratory pump; and (3) by inverting the body.

The muscles of the lower legs and feet play a crucial role in venous return (Masterson et al. 2006). Contraction of these muscles compresses the underlying veins and promotes the flow of blood back toward the heart. This is often referred to as the *skeletal muscle pump*. The calf muscles, situated at the back of the lower leg, are the largest muscles in this region and can be thought of as the heart of the legs because of the considerable part that they play in this process. When we stand for a long period without moving much, our feet can start to feel heavy as venous return slows down. One of the reasons you are encouraged to move around during a long flight is to promote this heart-of-the legs action to prevent deep vein thrombosis, which we look at in more detail later in the chapter. The practice of asanas in yoga can improve the strength and tone of the muscles in our lower legs and feet and may increase venous return as a result (Parshad, Richards, and Asnani 2011).

Respiratory activity also has a huge impact on venous return. In fact, a study by Miller and colleagues (2005) concluded that respiratory muscle pressure production is the predominant factor modulating venous return from the lower limb. This respiratory activity can be thought of as the respiratory pump. The mechanics of breathing can also affect the diameter of the venae cavae and cardiac chambers, which both directly and indirectly affect venous return. Byeon and colleagues (2012) reported that diaphragmatic breathing increases the efficiency of venous return and that the effect is maximized during slow respiration. (We discuss diaphragmatic breathing in more detail in chapter 3 on the respiratory system.) Intrathoracic pressure becomes progressively more negative by deep and slow inspiration during pranayama, and this also increases venous return (Parshad, Richards, and Asnani 2011). Dick and colleagues (2014) stated that slow breathing toward a rate of six breaths per minute results in increased venous return. The many yoga practices that focus on deep and slow diaphragmatic breathing can therefore increase venous return.

Inverting the body causes a transient increase in venous return (Haennel et al. 1988). Incorporating yoga asanas such as Supported Bridge Pose (Setu Bandha Sarvangasana) or Legs-Up-the-Wall (Viparita Karani) into a yoga practice can have a significant impact on venous return.

reasonable to believe that the blood vessels simply relax and therefore naturally dilate. This vasodilation may be direct, due to inadequate oxygen to sustain smooth muscle contraction, or indirect, via the production of vasodilator metabolites.

In reality, it is fair to imagine that a combination of the two mechanisms is involved.

Extrinsic Autoregulatory Processes

The autonomic nervous system and endocrine system both play a big role in the regulation of local blood flow. We look at this in more detail in the chapters on the nervous and endocrine systems.

RESTING HEART RATE AND CARDIAC OUTPUT

Resting heart rate (RHR) is simply a measure of how many times the heart beats in one minute when the body is at rest, which is normally in the range of 60 to 100 beats per minute (bpm). In a retrospective, longitudinal cohort study of over 90,000 adults (Quer et al. 2020), the mean daily RHR of the subjects was 65 bpm, with a range of 40 to 109 bpm among all individuals. The mean RHR differed significantly by age, sex, body mass index, and average sleep duration. Time of year variations were also noted, with a minimum in July and maximum in January. For most subjects, RHR

 ## Heavy Sweating During Hot Yoga Detoxes the Body

Detoxification is the physiological processes through which the body identifies, neutralizes, and eliminates toxic substances and metabolic byproducts. Detoxification is an essential part of homeostasis, and our bodies naturally possess the capacity to perform these processes very effectively. Without an effective detoxification system, we would be very unwell.

Sweat glands are often perceived to play an important excretory function, similar to that of the kidneys. However, in a comprehensive review, Baker (2019) concluded that the role of the sweat glands in eliminating waste products and toxicants from the body seems to be minor compared with other avenues of breakdown (liver) and excretion (kidneys and gastrointestinal tract). Studies suggesting a larger role of sweat glands in clearing waste products or toxicants from the body (e.g., concentrations in sweat greater than that of blood) may be an artifact of methodological issues rather than evidence for selective transport. Nevertheless, studies have shown that perspiration plays a role in skin hydration and microbial defense (Schröder and Harder 2006; Watabe et al. 2013).

So, it appears that heavy sweating in a hot yoga class or in a sauna might not help us to relinquish all those perceived toxins after all. However, practicing yoga as part of a healthy lifestyle will allow our kidneys, liver, and gastrointestinal tracts to continue to work optimally.

remained relatively stable over the short term, but 20 percent experienced at least one week in which their RHR fluctuated by 10 bpm or more. It is also interesting to note that our RHR increases up to three days before we have symptoms of a common cold or other infection. It is suggested that the heart rate typically rises by 10 beats for each Fahrenheit degree of fever (Tanner 1951). This might be one of the factors we inherently sense when we say, "I think I'm coming down with something."

Stroke volume is the volume of blood ejected from the heart per beat. At rest, the average stroke volume is approximately 70 milliliters. The Frank-Starling mechanism describes the ability of the heart to change its force of contraction and therefore stroke volume in response to changes in venous return. Cardiac output is the volume of blood ejected from the heart per minute and is measured by multiplying the stroke volume by the heart rate. The average cardiac output at rest is approximately five liters, and since adult females average four to five liters and adult males typically average about five to six liters of blood, we therefore circulate our entire volume of blood around the body every minute. Over one year, 2.6 million gallons (10 million L) of blood is sent through roughly 60,000 miles (96,560 km) of blood vessels. The heart, being a muscle, also grows in response to exercise. An increase in left ventricular mass increases cardiac output. Therefore, an elite athlete will have a very high cardiac output in relation to a sedentary individual (Fagard 2003).

HEART RATE VARIABILITY

Heart rate variability (HRV) is the variance in time between the beats of one's heart, and measuring HRV is a noninvasive way to identify autonomic nervous system imbalances. If a person's system is in a state of sympathetic activation, the variation between subsequent heartbeats is low. If one is in a state of parasympathetic activation, the variation between beats is high. In other words, the healthier the autonomic nervous system, the faster you are able to switch gears, showing more resilience and flexibility in your physiology.

Research has shown a relationship between low HRV and worsening depression or anxiety (Schiweck et al. 2019). A low HRV is even associated with an increased risk of death and cardiovascular disease (Buccelletti et al. 2009; Tsuji et al. 1994).

In a systematic review and meta-analysis of randomized clinical trials, Posadzki and colleagues (2015) concluded that there is no convincing evidence for the effectiveness of yoga in modulating HRV. They recommended that future investigations in this area should attempt to overcome the multiple methodological weaknesses of the previous research. However, a review by Tyagi and Cohen (2016) found that yoga can affect cardiac autonomic regulation with increased HRV and vagal dominance during yoga practices. Regular yoga practitioners were also found to have increased vagal tone at rest compared to nonyoga practitioners. (For more information about vagal tone, see page 51 in chapter 2 on the nervous system.) However, the authors concluded that it is premature to draw any firm conclusions about yoga and HRV as most studies were of poor quality, with small sample sizes and insufficient reporting of study design and statistical methods. While more research is needed, this study has shown that there is a possibility of yoga positively affecting HRV.

BLOOD PRESSURE

Blood pressure (BP) is the amount of pressure exerted on artery walls during contraction and relaxation of the heart. This pressure is essential for blood to travel throughout the body. BP readings are measured in millimeters of mercury (mm Hg) and are written as systolic pressure (the force the blood is exerting against the artery walls when the heart beats) over diastolic pressure (the force the blood is exerting against the artery walls while the heart is resting between beats). Therefore, a BP reading of 120/80 mm Hg (or verbalized as 120 over 80) tells us that the systolic pressure is 120 mm Hg and the diastolic pressure is 80 mm Hg. A normal BP reading is typically defined as less than 120/80 mm Hg.

Our bodies have a sophisticated way of constantly monitoring BP with specialized mechanoreceptors called *baroreceptors* (figure 4.5), which then provide information about BP to the autonomic nervous system. There are two types of baroreceptors: high-pressure arterial baroreceptors and low-pressure volume receptors. Both types of baroreceptors are stimulated by the stretching of vessel walls. Arterial baroreceptors are located within the carotid sinuses, which are at the base of the internal carotid arteries in the neck and the arch of the aorta. Low-pressure volume receptors, also known as *cardiopulmonary receptors*, are located within the atria, ventricles, and pulmonary vasculature (Al-Khazraji and Shoemaker 2018). The arterial baroreceptors are the terminals of afferent fibers that run in the glossopharyngeal and vagus nerves. Rapid decreases in BP result in decreased stretching of the artery wall and a decreased rate of impulse firing from the baroreceptors. This ultimately results in increased cardiac output and vasoconstriction that cause the BP to increase. The opposite is found to be true if BP rises rapidly.

Additional sensory receptors called *chemoreceptors* sense changes in our blood chemistry (oxygen, carbon dioxide, and pH levels) and are located in the same areas as baroreceptors. Low oxygen levels and high carbon dioxide levels cause the

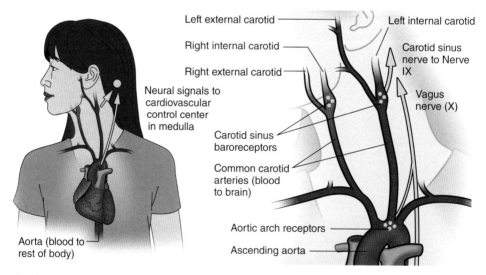

FIGURE 4.5   Location of baroreceptors.

chemoreceptors to send signals to the brain so that our respiratory rate and heart rate increase. High oxygen and low carbon dioxide have the opposite effect.

CONDITIONS OF THE CARDIOVASCULAR SYSTEM

We will now explore some of the main conditions that affect the cardiovascular system and discuss the role that yoga can play in improving these conditions.

Hypertension

High BP, known as *hypertension*, is when blood pressure is consistently too high. The primary way that hypertension causes harm is by increasing the workload of the heart and blood vessels, making them work harder and less efficiently. Over time, the force and friction of hypertension damages the delicate endothelium of the arteries. In turn, this can signify the start of a process called *atherosclerosis* (explored later in the section on cardiovascular disease).

Essential hypertension, a rise in blood pressure of undetermined cause, includes 90 percent of all hypertensive cases (Messerli, Williams, and Ritz 2007). Strong evidence points to a causal link between a chronically high salt intake and the development of hypertension (Meneton et al. 2005). Additional factors, such as obesity, diabetes, aging, emotional stress, sedentary lifestyle, and low potassium intake may increase the probability of developing hypertension (Takahashi et al. 2011).

Previous guidelines set the BP threshold at 140/90 mm Hg for people younger than 65 years of age and 150/80 mm Hg for those ages 65 and older. In 2017, new guidelines lowered the numbers for the diagnosis of hypertension to 130/80 mm Hg and higher for all adults. These new guidelines stem from the results of the Systolic Blood Pressure Intervention Trial (Whelton et al. 2018), which studied more than 9,000 adults aged 50 and older who had systolic blood pressure of 130 mm Hg or higher and at least one risk factor for cardiovascular disease. The study's aim was to find out whether treating blood pressure to lower the systolic number to 120 mm Hg or less was superior to the standard target of 140 mm Hg or less. The results found that aiming for a systolic pressure of no more than 120 mm Hg significantly reduced the chance of heart attacks, heart failure, or stroke over a three-year period.

A reading of 130/80 mm Hg and higher is considered stage 1 hypertension, while a reading of 140/90 mm Hg or higher is considered stage 2 hypertension. Anything higher than 180/120 mm Hg is termed a *hypertensive crisis*.

Current estimates suggest that over 76 million U.S. adults have hypertension (Roger et al. 2012) and that BP is well controlled in less than 50 percent of these individuals (Gillespie et al. 2011). According to a report by Kearney and colleagues (2005), the total number of adults with hypertension is predicted to increase to 1.56 billion worldwide by 2025.

Yoga students with hypertension that is well controlled can typically practice in the same way as someone with normal BP. If you have uncontrolled hypertension, it is important to get the go-ahead by your health care practitioner before practicing yoga. Students with uncontrolled high blood pressure need to be particularly mindful when

Can Yoga Lower Your Blood Pressure?

A systematic review and meta-analysis by Hagins and colleagues (2013) reported that yoga can be preliminarily recommended as an effective intervention for reducing BP. It is worth noting, however, that most published yoga and BP studies were not randomized, had inadequately described yoga or control programs, did not collect information on other lifestyle factors, and did not use standardized, reliable outcome measures. Additional rigorous controlled trials are warranted to further investigate this benefit of yoga.

The precise mechanisms by which yoga practices affect blood pressure currently remain unclear, although it has been proposed that yoga practices lower blood pressure by improving the sensitivity of baroreceptors and chemoreceptors, increasing vagal tone, and decreasing sympathetic nervous system drive (see page 45 for a full description) (figure 4.6). Small changes in BP are therefore detected sooner and controlled more quickly and effectively. In a small study by Vijayalakshmi and colleagues (2004), the authors concluded that yoga optimized the sympathetic response to stressful stimuli and restored the autonomic regulatory reflex mechanisms in hypertensive patients. Another small study by Selvamurthy and colleagues (1998) concluded that yoga training resulted in an improvement of baroreceptor sensitivity. Slow breathing, which is a very significant element of many yogic practices, has been shown to improve arterial baroreflex sensitivity and decrease BP in essential hypertension (Bernardia et al. 2001; Joseph et al. 2005). In a randomized controlled trial by Schneider and colleagues (1995), Transcendental Meditation more significantly reduced systolic and diastolic BP values than a control program. This tells us that the meditative element of yogic practices may also be playing a very significant role in lowering BP. Studies have shown that the medial prefrontal cortex and anterior cingulate gyrus are activated during attention-focusing tasks (Posner and Petersen 1990), and there is also evidence suggesting that a specific relationship between frontal lobe activity and BP exists (Williamson, McColl, and Mathews 2004). It is therefore fair to conclude that the focus that is inherently incorporated into all yogic practices is also significant here. While there is widespread belief that supine and inverted postures in yoga specifically stimulate the baroreceptor reflex, there is surprisingly very little research on this topic. A few studies several decades ago began to explore this (Cole 1989; Razin 1977; Tai and Colaco 1981) but it is challenging to find more recent literature that expands upon this topic.

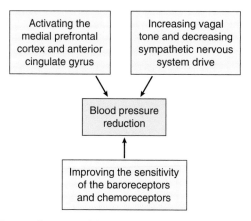

FIGURE 4.6   The effects of yoga on blood pressure.

it comes to inversions, because elevating the heart above the head and the torso and legs above the heart raises BP. A partially inverted posture like Downward Facing Dog (Adho Mukha Svanasana), where the heart is only slightly above the head and the legs are not elevated, may only slightly increase BP. There are many modifications for Downward Facing Dog that can be adopted, particularly if the student feels too much pressure in their head or feels short of breath. Supported Bridge Pose (Setu Bandha Sarvangasana), lying on bolsters with legs horizontal and feet at hip level, increases BP a little more. We will explore these yoga asanas in more detail later in the book. Students should always be given the option to come out of the asana if they feel uncomfortable in any way. Full inversions including Shoulder Stand (Salamba Sarvangasana) and Headstand (Sirsasana) should be avoided when a student has uncontrolled high blood pressure.

Hypotension

A BP reading that is less than 90/60 mm Hg is generally considered to be low BP or hypotension. Symptoms related to hypotension include dizziness or lightheadedness and nausea. If arterial pressure falls appreciably on standing, this is termed *orthostatic* or *postural hypotension*. This fall in arterial pressure can reduce cerebral blood flow to the point where a person might experience syncope (fainting). Low blood pressure after eating is termed postprandial hypotension. It is also common for BP to lower significantly during the first 24 weeks of pregnancy.

For yoga practitioners with hypotension, it is a good idea to take plenty of time to rise back up to vertical after folding forward. Raising the hands above the head causes BP to rise and can help manage symptoms of dizziness and lightheadedness. It is not uncommon to feel dizzy after raising the head after practicing backbends such as Camel Pose (Ustrasana). There is always the option to refrain from extending the neck in the first place or taking some extra time to slowly transition the head back to vertical.

Tachycardia and Bradycardia

Tachycardia is generally considered to be an RHR over 100 bpm. Tachycardia is considered a risk factor for cardiovascular morbidity and mortality in healthy people as well as those with cardiac diseases (Palatini 2011). This relationship has been found to be independent of age, sex, cholesterol levels, or blood pressure values in adults (Cooney et al. 2010).

It is widely recognized that improving fitness levels can lower one's RHR. A low RHR, known as *bradycardia*, is common in athletes (Jensen-Urstad et al. 1997), meaning that their hearts are working very efficiently at rest. Having a low RHR also means that a person will have a larger range before reaching their maximal heart rate, which can be a distinct advantage to an athlete. Boyett and colleagues (2013) suggested that this is a consequence of pacemaker cell alterations and intrinsic heart rate reduction. In a systematic review and meta-analysis, Reimers, Knapp, and Reimers (2018) concluded that exercise, especially endurance training and yoga, decreases RHR. Although the authors also suggested that this decrease of RHR may contribute to increasing life expectancy, this was not investigated in their meta-analysis and should therefore be a topic of further studies.

Anemia

Anemia is a decrease in the number of erythrocytes or a less-than-normal quantity of hemoglobin in the blood and is most commonly the result of iron deficiency, vitamin deficiency, or inflammation. People with anemia usually report feelings of weakness, fatigue, general malaise, and sometimes poor concentration.

A randomized controlled trial by Sharma and colleagues (2014) involved 23 female individuals with anemia. Individuals in the experimental group were progressively introduced to yogic practices, including asana, pranayama, meditation, and relaxation techniques, over a period of 30 days for around 70 minutes each day. The control group was not exposed to any yogic practices. The experimental group showed significantly improved levels of erythrocytes and hemoglobin. The control group did not show significant improvement in hemoglobin, and while this group did show some improvement in erythrocyte count, there was less improvement than in the experimental group. The authors concluded that yogic practices can be used efficiently to improve the hemoglobin count in individuals with anemia. The small sample size was the main limitation of this study.

Broken Heart Syndrome

Broken heart syndrome, or takotsubo cardiomyopathy, is a condition caused by extreme stress from such events as the death of a loved one, an emotional breakup, or loss of income. The recognized effects on the heart include congestive heart failure due to a profound weakening of the myocardium. The exact cause of this condition is unknown. While many patients survive the initial acute event with treatment to restore normal function, there is a strong correlation with death. A study by Spreeuw and Owadally (2013) revealed that within one year of the death of a loved one, women were more than twice as likely to die, and males were six times as likely to die as would otherwise be expected.

While there is currently no research that directly looks at the effect yoga has on such a syndrome, a study by Norcliffe-Kaufmann and colleagues (2016) concluded that women with a previous episode of broken heart syndrome had excessive sympathetic responsiveness and reduced parasympathetic modulation of heart rate, and this is where yoga could play a significant role.

This is a particularly eloquent quote from Judith Hanson Lasater (2017) about yoga and compassionate dying:

> *The practice of yoga is not a strategy for avoiding pain, even the pain we feel when we think about the inevitability of death; it is a way of confronting the issue and the pain directly. In the yoga tradition, deeply acknowledging the reality of death is said to be a source of freedom. By accepting our mortality, we can free ourselves from the bondage of avidya (ignorance). When we acknowledge death as inevitable instead of being blinded by our fear of it, everything else just comes into clearer focus, including the preciousness of each moment of life. (paragraph 7)*

Can Yoga Lower Your Risk of CVD?

A systematic review and meta-analysis by Cramer and colleagues (2014) revealed evidence for clinically important effects of yoga on most biological CVD risk factors. The authors included 44 randomized controlled trials with a total of 3,168 participants in the analysis. Despite methodological drawbacks of the included studies, the authors state that yoga can be considered a supporting intervention for the general population and for patients with increased risk of CVD. Interestingly, they reported that exactly 12 weeks of intervention duration seems to be more effective than shorter or longer interventions.

Cardiovascular Disease

Cardiovascular disease (CVD) is the leading cause of death in the United States (Heron 2019), with coronary artery disease being the most common type of CVD in the United States. It is sometimes called *coronary heart disease* or *ischemic heart disease*. Coronary artery disease is caused by plaque buildup in the walls of the coronary arteries. This plaque is made up of cholesterol deposits, and buildup of these deposits causes the lumen of the arteries to narrow over time, partially or totally blocking the blood flow in a process called *atherosclerosis*. The plaque can become unstable and rupture, often leading to additional clot formation.

Hypertension, high cholesterol, and smoking are all risk factors that can lead to CVD and stroke. In 2009 to 2010, approximately 46.5 percent of U.S. adults aged 20 and over had at least one of these three risk factors (Fryar, Chen, and Li 2012). Several other medical conditions and lifestyle choices can also put people at a higher risk for heart disease, including diabetes, obesity, unhealthy diet, physical inactivity, and excessive alcohol use. Chronic stress exposure has also been found to have a strong link to CVD (Rosengren et al. 2004) and to CVD risk factors (Bhavanani 2016).

Deep Vein Thrombosis

Deep vein thrombosis (DVT) occurs when a blood clot forms in the deep leg veins, particularly in the region of the calf muscles (figure 4.7). Pulmonary embolism, a potentially life-threatening complication, is caused by the detachment of a clot that travels to the lungs. Nonspecific signs of DVT may include pain, swelling, redness, warmness, and engorged superficial veins.

DVT is predominantly a disease of the elderly with an incidence that rises markedly with age (Silverstein et al. 1998). Pregnancy has also been shown to be a risk factor (Bates and Ginsberg 2001), and the approximate risk for DVT following general surgery procedures is 15 to 40 percent. This risk of DVT has also been found to change depending on ethnicity with African American people being in the highest risk group for first-time DVT while Hispanic people's risk is about half that of Caucasians (Keenan and White 2007). Prolonged immobilization, including being on a long-haul

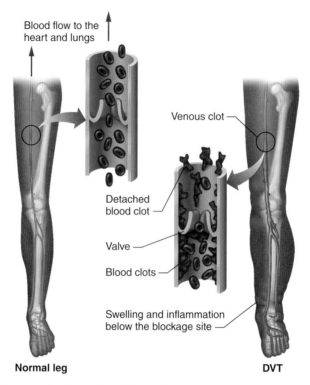

Blood flow to the heart and lungs

Venous clot

Detached blood clot

Valve

Blood clots

Swelling and inflammation below the blockage site

Normal leg   DVT

FIGURE 4.7   Deep vein thrombosis (DVT).

flight, is another risk factor (Gavish and Brenner 2011). When we are on a long flight, it is recommended to do some simple leg exercises to help prevent pooling of blood in our lower legs and therefore the formation of a clot.

While there is little to no research that specifically looks at how yoga can affect the risk of DVT, since yoga improves venous return and improves ankle mobility, it is fair to imagine that a regular yoga practice may lower the risk of DVT. It is important to note that yoga students with undiagnosed leg swelling should be encouraged to seek medical attention before practicing yoga. Exercising while having a DVT event may increase the risk of a pulmonary embolism and should be performed under the guidance of a medical professional.

Varicose Veins

Varicose veins (figure 4.8) are a common manifestation of reduced venous return in the lower limb. They appear as dilated, elongated, or twisted superficial veins. As a result of varicose veins, raised venous pressure in the lower leg can result in skin changes such as hyperpigmentation and hardening with eventual tearing of the veins,

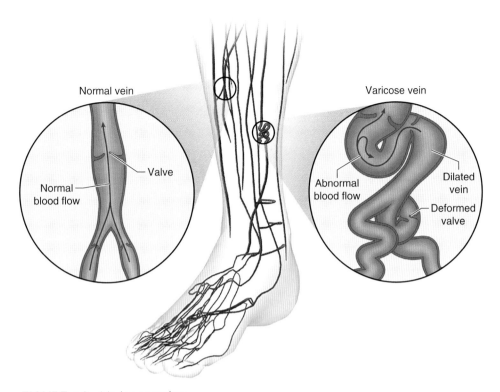

FIGURE 4.8 Varicose veins.

often referred to clinically as *chronic venous insufficiency*. Approximately one-third of men and women aged 18 to 64 have varicose veins (Evans et al. 1999), the prevalence of which increases with age. The evidence to support any association between the development of varices and lifestyle factors, such as prolonged standing, is lacking.

A study by Kravtsov and colleagues (2016) found that strengthening the muscular component of the musculovenous pump led to an improvement of the clinical course of varicose vein disease. Therefore, yoga practices that focus on improving the strength and tone of the muscles in the lower leg may help to improve the signs and symptoms of varicose veins. It is worth noting that prolonged standing can sometimes worsen varicose vein symptoms for some people. In a yoga class, one can always sit or lie down for a few moments during a long sequence of standing poses to reduce the pressure in the lower legs. Earlier we discussed how incorporating asanas such as Supported Bridge Pose (Setu Bandha Sarvangasana) or Legs-Up-the-Wall (Viparita Karani) into a yoga practice may have a significant impact on venous return, and this may be beneficial to many students who have varicose veins.

TRY IT YOURSELF: Dirga Pranayama

Dirga pranayama, often known as *three-part breath*, can be a very calming and grounding practice that you can perform in any comfortable seated position or lying down. Close your eyes if you like, or just soften your gaze and focus on a fixed object in front of you. Release your jaw, let your lips and teeth gently part, and allow your tongue to fall away from the roof of your mouth. Start by taking a couple of gentle breaths in and out through your nose. During your next inhalation, focus on the gentle expansion of your abdomen and then slowly and fully exhale. During the following inhalation, focus on the abdominal expansion but also allow your lower chest to gently expand. Then fully exhale. On the third inhalation, allow your abdomen to expand, your lower chest to rise, and your upper chest to expand. Slowly and fully exhale. For the next few inhalations, focus on your abdomen, your lower chest, and your upper chest expanding. When you are ready, come back to a more natural breath with less effort and notice how you are feeling after this short pranayama practice.

CONCLUSION

The cardiovascular system is a fascinating and integral part of the human body. Yoga is often a heart-centered practice, and it is reassuring to know that science supports the view that we are doing something genuinely beneficial for our cardiovascular system when we practice yoga.

Chapter 5

LYMPHATIC AND IMMUNE SYSTEMS

There is widespread belief among yoga teachers and practitioners that practicing yoga boosts the immune and lymphatic systems. In this chapter, we explore these two vital systems and discover if this claim is supported by research.

The lymphatic and immune systems are so inextricably linked that it is almost impossible to thoroughly discuss one without including the other. While the lymphatic vasculature is not formally considered a part of the immune system, it is critical to immunity with one of its major roles being the coordination of the trafficking of antigen and immune cells. Evidence that the lymphatic system plays additional roles in immunity is also emerging.

LYMPHATIC SYSTEM

The lymphatic system has been somewhat neglected in the past by both the scientific and medical communities because of its vagueness in structure and function. The amount of available information on this fascinating system is scarce, particularly compared to such systems as the cardiovascular system, but it is no less significant than any other bodily system.

The lymphatic system is a linear, one-way system of vessels, cells, and organs that runs parallel to and entwines the blood circulatory system. It carries excess fluids from the tissues of the body to the bloodstream and plays an integral role in the immune functions of the body. It also plays an important role in the process of inflammation and is involved in the transport of dietary fats and fat-soluble vitamins absorbed in the gut.

Lymph and the Lymph Vessels

The leakage of plasma from the capillaries of the cardiovascular system creates interstitial fluid that bathes the surrounding cells, supplying them with nutrients and oxygen and collecting carbon dioxide and other waste products. It has been estimated that the total plasma volume of the human body (approximately three liters) leaks from the blood circulation every nine hours, and while some interstitial fluid is reabsorbed directly by the blood vessels, the majority of this fluid is transported back to systemic circulation through the lymphatic system (Levick and Michel 2010). Once the interstitial fluid enters the lymphatic system, it becomes lymph.

Lymphatic capillaries are interwoven with the arterioles and venules of the cardiovascular system and are lined with a single layer of partly overlapping lymphatic endothelial cells, which mechanically function as primary valves that unidirectionally control lymph fluid drainage. The membrane around the lymphatic capillary is rather sparse and does little in the way of a filtration barrier (Pflicke and Sixt 2009), allowing for the nonselective uptake of interstitial contents including large macromolecules such as pathogens and migrating cells.

The lymphatic capillaries feed into larger and larger lymphatic vessels that are similar to veins in terms of their three-tunic structure. These vessels carry the lymph to a series of lymph nodes (figure 5.1). In general, lymphatic vessels of the tissues of the skin, known as *superficial lymphatics*, follow the same routes as veins, whereas the deep lymphatic vessels of the viscera generally follow the paths of arteries. The superficial and deep lymphatic vessels eventually merge to form lymph trunks. Four pairs of lymph trunks are distributed laterally around the center of the body, along with an unpaired intestinal trunk. The lymph trunks then converge into the two lymph ducts: the right lymph duct and the thoracic duct. Interestingly, the overall drainage of the body is asymmetrical: The right lymph duct receives lymph from only the upper right side of the body while the lymph from the rest of the body enters the bloodstream through the thoracic duct via all the remaining lymph trunks. The right duct drains into the right subclavian vein while the thoracic duct drains into the left subclavian vein, both at the junction between the respective subclavian vein and the jugular vein. The subclavian and jugular veins are all deep neck veins. The lymphovenous valves at the junction of the ducts and subclavian veins are crucial to preventing blood, which is at a higher pressure in veins than the pressure in the thoracic duct, from backing up into the thoracic duct and reaching upstream lymph nodes and even peripheral tissues like the intestines (Hess et al. 2014).

Lymphatic vessels have been identified in organs where they were previously not thought to exist, including the eye, where they are involved in intraocular pressure regulation, and in the central nervous system, where they drain cerebral interstitial fluid, cerebrospinal fluid, macromolecules, and immune cells (Aspelund et al. 2016). The only organ that is not believed to have lymphatic vessels is the bone marrow (Edwards et al. 2008).

Lymph Flow

Unlike blood, lymph does not have a pump like the heart to help keep it circulating, although the presence of one-way valves within the lymph vessels does prevent backflow of lymph. These one-way valves are located close to one another, and each one causes a bulge in the lymphatic vessel, giving the vessels a beaded appearance.

Since the late 1900s, knowledge about the physiology of the lymphatic system has grown more complete. However, there is still no generally accepted model that adequately describes the mechanisms and regulation of lymph transport. Lymph flow is likely to be the result of a complex combination of both active and passive driving forces.

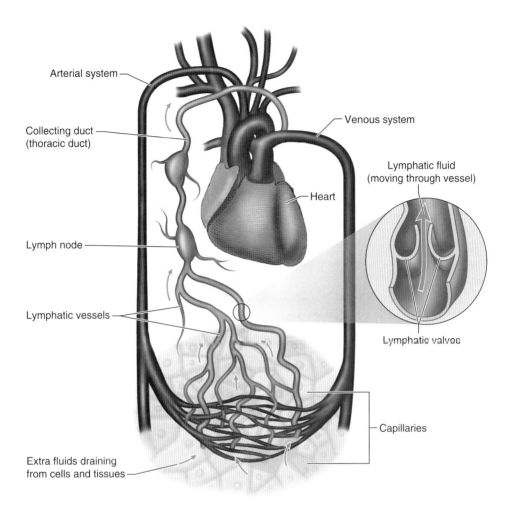

FIGURE 5.1 Lymphatic vessels running parallel to the cardiovascular system.

The intrinsic driving force, known as the *active pump*, is created by a lymphangion (figure 5.2), which is the section of a lymphatic vessel between two adjacent lymphatic valves (Gashev and Zawieja 2001). Lymphangions were first described by Mislin in 1961 and act like the ventricles of the heart.

Coordinated contractions of these sections are initiated by the pacemaker activity of smooth muscle cells and modulated by the pressure gradient across the vessel wall (Gashev 2002). It is important to note that our precise understanding of this process is somewhat limited, although it is believed that the contractions spread from one lymphangion to the next like a wave that causes contraction along both the length and width of the vessels (Margaris and Black 2012). The contractions can be compared to the rhythmical contraction of the digestive system, known as *peristalsis*.

Extrinsic driving forces, known as the *passive pump*, include lymph formation, arterial pulsations, skeletal muscle contractions, fluctuations of central venous

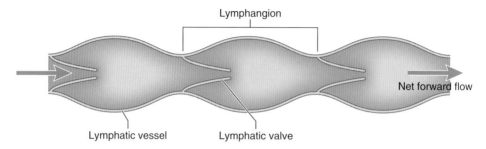

FIGURE 5.2 Lymphangion.

pressure, gastrointestinal peristalsis, and respiration. These forces produce passive hydrostatic gradients in the lymphatic network, which may effectively propel lymph (Gashev 2002). It is not clear whether extrinsic mechanisms can have a significant or even dominant role in the pumping of lymph. Engeset and colleagues (1977) suggested that, at rest, approximately one-third of lymph transport in the lower extremities results from compression by skeletal muscle contractions and two-thirds results from the active pumping of the collecting vessel network.

Exercise is thought to help increase lymph flow via muscle contraction around the lymphatics (Cheville et al. 2003). External forces such as massage have been shown to affect the filling of lymphatic capillaries rather than the pumping of the larger lymphatic vessels (Auckland 2005).

Like other bodily systems, the lymphatic system is under the control of the nervous and endocrine systems, matching lymphatic pumping to the physiological activities of other parts of the body.

Lymph Nodes

Lymph nodes (figure 5.3) are small, kidney-shaped glands positioned throughout the lymphatic system. There may be anywhere from 500 to 600 lymph nodes in the human body and while they are concentrated around the neck, groin, armpit, and behind the knees, lymph nodes are actually present throughout the body. They are surrounded by a dense connective tissue capsule and have an inner cortex and medulla, which contain blood vessels and many cells of the immune system. They can be thought of as filters of the lymph because of their role in removing dead or damaged cells, large protein molecules, toxins, and pathogens from the lymph.

Lymph nodes receive lymph via multiple afferent vessels, and filtered lymph then leaves via one or two efferent vessels. Lymph nodes substantially concentrate lymph through the absorption of water so that the protein concentration of lymph has essentially doubled by the time it exits a lymph node (Adair et al. 1982). If lymph nodes become swollen, painful, or hard, it can be a sign of an active defense reaction and is called *lymphadenopathy*. Perhaps you have noticed how the lymph nodes under your jaw become swollen when you have a cold. We discuss the role that the lymph nodes play in the immune system in more detail later in the chapter.

Diaphragmatic Breathing and Postural Changes in Yoga Improve Lymphatic Drainage

If you type *yoga and lymphatic drainage* into any search engine, you will come across hundreds of articles confidently stating that yoga improves the functioning of the lymphatic system. Searching deeper to find evidence that backs up this claim proves much more challenging. There is a scarcity of scientific and medical information about the lymphatic system in general, and information about this system specifically related to yoga, respiration, or posture is even more scarce. While there is certainty about the impact of the phases of respiration on venous return, there is a great deal of uncertainty and many differing opinions about its impact on the lymphatic system (Piller et al. 2006). The same applies for the role that inverting the body in poses like Shoulder Stand (Salamba Sarvangasana) and Headstand (Sirsasana) plays on the lymphatic system. Some sources (Cemal, Pusic, and Mehrara 2011; Seki 1979) discuss how postural changes may affect the flow of lymph. However, while common sense tells us that inverting the body is likely to have some effect on improving lymphatic drainage, it is important to note that for most of us, our lymphatic system is already working very effectively and does not need to be improved.

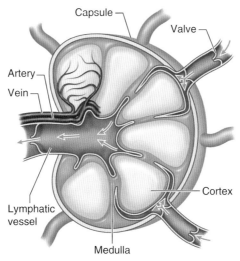

FIGURE 5.3   Lymph node.

Lymph and the Digestive System

In the small intestine, lymphatic capillaries called *lacteals* are critical for the transport of dietary fats and fat-soluble vitamins to the bloodstream. These nutrients enter the lacteals to form a milky fluid called *chyle*, which then travels through the lymphatic system and is eventually added to the bloodstream. In the fasting state, intestinal lymph production is thought to be relatively low but will increase significantly after a meal is eaten. It is suggested that lacteals contract through the activity of the surrounding smooth muscle cells regulated by the autonomic nervous system (Choe et al. 2015).

Role of the Lymphatic System in Inflammation

Inflammation is a defensive reaction against pathogens or irritants and is characterized by the five cardinal symptoms of redness, increased heat, swelling, pain, and impaired function. While acute inflammation is a crucial part of our immune process, there are many instances when persistent inflammation leads to serious health problems, for example, in rheumatoid arthritis or inflammatory bowel disease. The lymphatic vasculature plays a crucial role in regulating the inflammatory response by influencing drainage of interstitial fluid, inflammatory mediators, and leukocytes. Lymphatic vessels undergo pronounced enlargement in inflamed tissue and display increased leakiness, indicating reduced functionality. However, stimulation of the lymphatic vasculature has been shown to reduce inflammation severity in models of rheumatoid arthritis, skin inflammation, and inflammatory bowel disease. This beneficial effect of lymphatic activation in inflammation may represent a promising therapeutic approach in the setting of inflammatory pathologies (Schwager and Detmar 2019).

Role of the Lymphatic System in Cardiovascular Disease

Recent research has shed light on the involvement of the lymphatic system in cardiovascular diseases including atherosclerosis, hypertension, and myocardial infarction (Aspelund et al. 2016). The heart has an extensive lymphatic capillary network that ensures optimal cardiac function. As little as a 2.5 percent increase in tissue fluid content can lead to a 30 to 40 percent reduction in cardiac output (Dongaonkar et al. 2010). Cardiac lymphatic vessel growth, or lymphangiogenesis, occurs in response to injury of cardiac muscle (myocardial infarction), whereby induced lymphatic sprouting results in improved prognosis (Klotz et al. 2015). Cholesterol plays a very significant role in the development of atherosclerosis, the major underlying factor of cardiovascular disease. The removal and eventual excretion of excess cholesterol from peripheral tissues is termed *reverse cholesterol transport* (*RCT*). Research indicates that RCT depends critically on lymphatic vessels, and that the venous system is not enough to sustain RCT. Furthermore, inducing lymphangiogenesis could potentially constitute a strategy to enhance RCT (Lim et al. 2013).

TRY IT YOURSELF: Simple Breath Work

Start by finding a comfortable position, either sitting or lying down. Once you are comfortable, either close your eyes or soften your gaze and focus on a fixed point in front of you. Release your jaw, let your lips and teeth gently part, and allow your tongue to fall away from the roof of your mouth. Take a moment to scan your body with your mind, observing how you are feeling physically today. Try your best not to judge the sensations that you notice but simply observe. Start now to observe your breath for a few moments, noticing its quality. Is its rate slow or fast? Is it shallow or deep? Which parts of your body are moving as you inhale and exhale? Now begin to focus on gently inhaling and exhaling through your nose. Place your hands on your lower abdomen so that your little fingers rest into your hip creases. Begin to notice your abdomen gently expand as you inhale and gently fall again as you exhale. Each time your mind naturally begins to wander off, gently draw your focus back to the rise and fall of your abdomen. Begin now to focus your attention on the exhalations. Without straining, allow each exhalation to gently lengthen and then pause for a few moments at the end of the exhalation. Allow each inhalation to arise naturally without any effort and then focus once again on the exhalation. Repeat this practice for a few minutes and then when you are ready, begin to breath more naturally again with less effort. Reflect on how you are feeling after this short pranayama practice.

CONDITIONS OF THE LYMPHATIC SYSTEM

We will now explore two of the main conditions that affect the lymphatic system and discuss the potential role that yoga can play in managing these conditions.

Lymphedema

Lymphedema is the excess accumulation of interstitial fluid in the tissue spaces. It can increase the risk of skin infection (cellulitis) and infection of the lymph vessels (lymphangitis), make movement very challenging, and it can lead to fatigue, sleeping problems, concerns about body image, and significantly decreased quality of life. Lymphedemas are classified as inherited (primary) or acquired (secondary). Primary lymphedemas result from defects in genes involved in lymphatic vessel development, while secondary lymphedemas arise from damage or physical obstruction of lymphatic vessels or lymph nodes. This could be the consequence of infection, trauma, surgery, transplantation, medication, or venous disease (Margaris and Black 2012). The surgical removal and radiotherapy of the breast and associated axillary lymph nodes results in lymphedema in 6 to 30 percent of patients (Petrek and Heelan 1998).

There is a notable disparity between the amount of research conducted for the treatment of lymphatic diseases and for cardiovascular diseases. In a systematic

Can Yoga Help to Improve Lymphedema?

A six-month study by Douglass and colleagues (2012) aimed to ascertain if regular yoga practice can impart any benefit in terms of lymphedema status, symptoms, or overall quality of life. They concluded that although there was no statistically significant difference from the control group, trends suggested a benefit in continuing yoga for reduction of lymphedema and justified further investigation by future larger, longer term investigations.

A systematic review by Wanchai and Armer (2020) reported that yoga is not shown to be an effective strategy for managing or preventing breast cancer–related lymphedema. The authors note that there were few related studies, that research methodology was of poor quality, and that sample sizes were small. Hopefully in the future we will see additional good-quality research being done to explore this important topic.

---

review and meta-analysis of randomized controlled trials, the authors concluded that current evidence does not support the use of manual lymphatic drainage in preventing or treating lymphedema (Huang et al. 2013). In another systematic review and meta-analysis (McNeely et al. 2011), the findings supported the use of compression garments and compression bandaging for reducing lymphedema volume in upper and lower extremity cancer-related lymphedema. Specific to breast cancer, a benefit was found from the addition of manual lymph drainage massage to compression therapy for upper extremity lymphedema volume.

Lymphoma

Lymphoma is the name for a group of blood cancers that develop in the lymphatic system. The two main types are Hodgkin lymphoma and non-Hodgkin lymphoma. Hodgkin lymphoma is one of the most curable forms of cancer and is named after Dr. Thomas Hodgkin who, in 1832, described several cases of people with symptoms of a cancer involving the lymph nodes. Non-Hodgkin lymphoma is a type of cancer that generally develops in the lymph nodes and lymphatic tissue found in organs such as the stomach, intestines, or skin. In some cases, it can also involve bone marrow and blood. It arises from lymphocytes, and lymphoma cells may develop in just one place or in many sites in the body.

IMMUNE SYSTEM

Microbes are tiny organisms that are found everywhere in our environment. The most common types are bacteria, viruses, fungi, and protozoa. The vast majority of microbes are harmless to us, and many play essential roles in plant, animal, and human health. On average, we are made up of about 30 trillion cells but carry a similar number of bacteria, mostly in the gut (Sender, Fuchs, and Milo 2016). A pathogen is defined as a microbe that can cause disease or death to its host. Only one in a billion microbial species is a human pathogen. Indeed, approximately 1,400 human pathogens have been described, whereas there are an estimated 1 trillion microbial species on Earth, the vast majority of which remain uncharacterized (Balloux and van Dorp 2017).

Can Yoga Help to Improve Quality of Life for People Living With Cancer?

Buffart and colleagues (2012) completed a systematic review and meta-analysis of randomized controlled trials that explored the physical and psychosocial benefits of yoga in cancer patients and survivors. Sixteen publications of 13 randomized controlled trials met their inclusion criteria, of which one included patients with lymphomas, and the others focused on patients with breast cancer. This review found that yoga had large beneficial effects on distress, anxiety, and depression; moderate beneficial effects on fatigue, general health-related quality of life, emotional function, and social function; and a small and insignificant effect on sleep and physical function and symptoms. A systematic review by Sharma, Haider, and Knowlden (2013) looked at yoga as an alternative and complementary treatment for the psychological and physical factors associated with cancer. Thirteen studies met the inclusion criteria, with six of them using a randomized controlled design. The authors concluded that the evidence for the efficacy of yoga as an alternative and complementary treatment for cancer is mixed, although generally positive. They stated that the limitations of the reviewed interventions included a mixed use of instruments, weak quantitative designs, small sample sizes, and a lack of theory-based studies.

In a randomized control trial, Cohen and colleagues (2004) looked at the effects of a yoga intervention on psychological adjustment and sleep in patients with lymphoma. The authors concluded that a yoga program is feasible for patients with cancer and that such a program significantly improves sleep-related outcomes. However, there were no significant differences between groups for the other outcomes. Clearly, yoga is not going to cure cancer. But what these studies show is that yoga can be a very beneficial additional tool for someone diagnosed with cancer, and it can significantly improve quality of life. While it is not a cure, that is still hugely important.

The immune system is the collection of cells and organs that communicate in complex ways to destroy or neutralize pathogens. It has evolved for the maintenance of homeostasis, as it is sophisticated enough to discriminate between foreign substances and self. However, when this specificity is affected, an autoimmune reaction or disease develops.

Active Versus Passive Immunity

Active immunity develops when exposure to a pathogen triggers the immune system to produce antibodies to that disease. Active immunity can be described as natural or vaccine induced. Natural immunity occurs when we are exposed to the pathogen through infection with the actual disease, while vaccine-induced immunity comes about through the introduction of an inactive or weakened form of the pathogen as a vaccination. If an immune person comes into contact with that disease in the future, their immune system will recognize it and immediately produce the antibodies needed to fight it. Active immunity can take several weeks to develop but is long lasting and sometimes lifelong.

Passive immunity is provided when a person is given antibodies to a disease rather than producing them through their own immune system. As fetuses, we acquire passive immunity from our mother through the placenta. Breast milk also contains antibodies, which means that babies who are breastfed have passive immunity for longer. A person can also obtain passive immunity through antibody-containing blood products, which may be given when immediate protection from a specific disease is needed. Passive immunity is immediate but lasts for only a few weeks or months.

Classification of White Blood Cells

White blood cells are the cells of the immune system and are known as *leukocytes* (figure 5.4). They can be split into two main categories: granulocytes and agranulocytes. Granulocytes are leukocytes that have small granules, and there are three types: neutrophils, eosinophils, and basophils. The granules contain proteins and enzymes that help to kill bacteria. Agranulocytes are leukocytes without granules. There are two types of agranulocytes: lymphocytes and monocytes. Lymphocytes can then be broken down into three types: T lymphocytes, B lymphocytes, and natural killer cells. Monocytes are the largest of the leukocytes and are three to four times the size of red blood cells.

Innate Immune Response

The innate immune response, often our first line of defense, defends the body against a pathogen in a nonspecific but rapid way. An example of innate immunity is the inflammatory immune response, which blocks the entry of pathogens through the skin, the respiratory tract, or the gastrointestinal tract. Physical barriers such as these, as well as the nasopharynx, eyelashes, and other body hair play a really important role here. Additional defense mechanisms such as mucus, bile, gastric acid, saliva, tears, and sweat are also incredibly significant.

There are many specific leukocytes that play vital roles in innate immunity:

- Neutrophils are the first leukocytes recruited to sites of acute inflammation where they engulf and ingest pathogens in a process called *phagocytosis*. They are the most common type of leukocytes, accounting for 55 to 70 percent of the total leukocyte count. Neutrophils are remarkably short lived with a circulating half-life of six to eight hours and are therefore produced at a rate of between 50 and 100 billion cells per day (Summers et al. 2010).

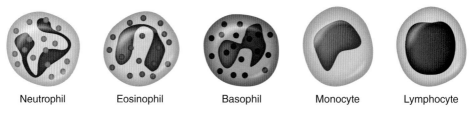

Neutrophil   Eosinophil   Basophil   Monocyte   Lymphocyte

FIGURE 5.4 Leukocyte delineation.

Exposure to Cold Weather Weakens the Immune System

We have surely all been warned, "Don't forget your coat or you'll catch a cold!" However, a review of the research studies on this topic by Castellani, Brenner, and Rhind (2003) concluded that moderate cold exposure has no detrimental effect on the human immune system. It is suggested that colds and flus are more prevalent in the winter because people spend more time indoors, in closer contact with other people who can pass on their germs, and viruses like influenza stay airborne longer when air is cold and less humid. A report by Xu and colleagues (2020) looking at COVID-19 transmission suggested that warmer temperature and moderate outdoor ultraviolet exposure may offer a modest reduction in transmission. Furthermore, cold exposure training can be beneficial to building adaptations that improve our tolerance to cold environments, a technique that climbers commonly use before venturing up a mountain. So, when you are out in cold weather, stay wrapped up for your comfort, but let go of the idea that by doing so you are somehow protecting your immune system.

- Eosinophils secrete a range of highly toxic proteins and free radicals that kill bacteria and parasites. They are found in many locations, including the thymus, lower gastrointestinal tract, ovaries, uterus, spleen, and lymph nodes.

- Basophils are the only circulating leukocytes that contain histamine, and they share many similarities with the mast cell.

- Mast cells derive from the bone marrow and do not fully mature until they are recruited into the tissue where they undergo their terminal differentiation. They are found in mucous membranes and connective tissues and are important for wound healing and defense against pathogens via the inflammatory response. When mast cells are activated, they release chemical molecules to create an inflammatory cascade. Mediators, such as histamine, cause blood vessels to dilate, increasing blood flow and cell trafficking to the area of infection.

- Natural killer cells are lymphocytes that do not attack pathogens directly but destroy infected host cells in order to stop the spread of an infection. Infected or compromised host cells can signal natural killer cells for destruction through the expression of specific receptors and antigen presentation. (Antigens are toxins or other foreign substances that induce an immune response in the body.)

- Monocytes are circulating precursor cells that differentiate into either macrophages or dendritic cells, which can be rapidly attracted to areas of infection by signal molecules of inflammation.

- Macrophages can move across the walls of capillary vessels, and their ability to roam outside of the circulatory system allows them to hunt pathogens with fewer limits. Macrophages can also release special proteins called *cytokines* in order to signal and recruit other cells to an area with pathogens.

- Dendritic cells are the most potent type of antigen-presenting cells. They are located in tissues and can contact external environments through the skin and through the inner mucosal lining of the nose, lungs, stomach, and intestines. Since dendritic cells are located in tissues that are common points for initial infection, they can identify threats and act as messengers for the rest of the immune system by antigen presentation. Dendritic cells are also responsible for the initiation of adaptive immune responses and hence function as the sentinels of the immune system.

Adaptive Immune Response

The slower but more specific and effective adaptive immune response involves many cell types and soluble factors but is primarily controlled by B and T lymphocytes.

B lymphocytes function primarily by producing antibodies, which are proteins that bind to a particular molecular component of a pathogen called an *antigen* and neutralize them. These activated B lymphocytes are known as *plasma cells*. The human immune system can generate billions of types of antibodies; this process is known as *humoral immunity*.

T lymphocytes are classified as helper T cells or cytotoxic T cells. Each of these cells develops its own T cell receptor (TCR) that is specific for a particular antigen.

Helper T cells have TCRs and special receptors on their surface called *CD4 receptors* that bind to antigen-presenting cells when their TCRs recognize the antigen being presented. Once bound, helper T cells release cytokines to stimulate a defense against that specific antigen.

Infected cells actively making viral proteins present pieces of those proteins on their surfaces. Cytotoxic T cells have TCRs and special receptors on their surface called *CD8 receptors*. When their TCRs match the viral antigen, they proceed to kill the infected cells.

Memory T cells are antigen-specific T lymphocytes that remain long after an infection has been eliminated. The memory T cells are quickly converted into large numbers of effector T lymphocytes upon reexposure to the specific invading antigen, thus providing a rapid response to past infection.

While B lymphocytes mature in red bone marrow, T lymphocytes mature in the thymus. Both B and T lymphocytes are found in many parts of the body, circulating in the bloodstream and lymph, and residing in secondary lymphoid organs, including the spleen and lymph nodes.

Immune Function and the Lymphatic System

Many studies indicate that lymphatic vessels are not mere passive conduits for immune cells but that they actively participate in influencing immune responses mediated by specific lymphocytes (Choi, Lee, and Hong 2012). Antibodies are also produced in the lymph nodes, which prevents recurrent infections from the same type of pathogen.

Primary Lymphoid Organs

The bone marrow and thymus gland are referred to as the *primary lymphoid organs*. Bone marrow (figure 5.5) is found in the central cavity of bone and is classified as

The Immune System Can Be Enhanced

There are many claims about different lifestyle interventions enhancing our immune system, but the concept of boosting immunity actually makes little sense scientifically. In fact, boosting the number of immune cells in your body is not necessarily a good thing. If your innate immune response was constantly stimulated, you would feel permanently unwell with a runny nose, fever, and lethargy. Inflammation has also been linked to depression (Haapakoski et al. 2016). Luckily there is no way to intentionally boost the innate immune system. It is also important to note that for the vast majority of us, the body already produces many more lymphocytes than it can possibly use, and the extra cells remove themselves through a natural process of cell death called *apoptosis*. On the whole, our immune system already does a remarkable job of defending us against disease-causing microorganisms without needing to be further enhanced. Although some lifestyle interventions have been found to alter some immune system components, currently there is no evidence that they actually boost immunity to the point where the person is better protected against infection and disease.

It is widely accepted that moderate exercise is good for us. Just like a balanced diet, exercise can contribute to general good health and therefore to maintaining a healthy immune system. Exercise promotes good blood circulation, which also helps the immune system to work optimally, and is one of the most widely studied behavioral interventions in terms of its immunomodulatory effects. Studies demonstrate an association between physical inactivity and low-grade systemic inflammation in healthy subjects, while regular exercise protects against diseases associated with chronic low-grade systemic inflammation (Petersen and Pedersen 2005). A study by Martin, Pence, and Woods (2009) suggested that moderate intensity exercise improves the immune response to respiratory viral infections. In a systematic review of the literature, Ploeger and colleagues (2009) investigated the effects of acute and chronic exercise on various inflammatory markers in patients with a chronic inflammatory disease. They reported that while training programs can reduce chronic inflammation in some patients, single bouts of exercise might elicit an aggravated inflammatory response. They suggested that the exercise training-induced response appears highly dependent on the type of disease; the severity of the disease; and the frequency, duration, and intensity of the exercise intervention. The authors also noted that the results of the review reveal a major gap in our knowledge regarding the evidence for safe but effective exercise for patients with a chronic inflammatory disease. A study by Haaland and colleagues (2008) also reported that the intensity of the exercise plays an important role. They suggested that strenuous exercise may cause acute immunologic changes (such as diminished natural killer cell activity), which may predispose to infection in certain individuals.

It is also widely recognized among the scientific community that people who are malnourished are more vulnerable to infectious diseases. However, whether the increased rate of disease is caused by the effect of malnutrition on the immune system is not certain. There are still relatively few studies on the effects of nutrition on the immune system of humans. While a systematic review by Rytter and colleagues (2014) concluded that the immunological alterations associated with malnutrition in children may contribute to increased mortality, the underlying mechanisms are still

(continued)

 **The Immune System Can Be Enhanced** *(continued)*

inadequately understood. The authors also noted that different types of malnutrition are associated with different immunological alterations. They suggested that better designed prospective studies are needed, based on current understanding of immunology and with state-of-the-art methods. In a systematic review and meta-analysis of randomized controlled trials looking at the role of multivitamins and mineral supplements in preventing infections in elderly people, the authors (El-Kadiki and Sutton 2005) concluded that the evidence for routine use of multivitamin and mineral supplements to reduce infections in elderly people is weak and conflicting. Only eight trials met their inclusion criteria, and owing to inconsistency in the outcomes reported, not all of the trials could be included in each meta-analysis. Much more good quality research is needed to give us a better understanding of the role that multivitamins and supplements play here. If you already have a balanced diet that meets the recommended amounts of nutrients, there is no need to take vitamin and mineral supplements, and doing so does not enhance your immune system. If you are malnourished in some way, then taking specific supplements may help you to reach recommended nutritional levels.

While it is important to reaffirm that moderate exercise and a balanced diet are good for us, it is also important to reiterate that there are currently no scientifically proven direct links between lifestyle interventions and *enhanced* immune function.

---

yellow or red. Yellow bone marrow consists of cells specialized for the storage of fat, while red bone marrow is the production site for erythrocytes, platelets, dendritic cells, and various lymphocyte subsets including B lymphocytes. In adults, red bone marrow persists in the ribs, clavicula, scapulae, pelvis, and vertebrae, as well as in the sternum and the proximal ends of the femurs. T lymphocytes are also known to migrate to the bone marrow from the periphery. Plasma cells are differentiated B lymphocytes capable of secreting antibodies and can survive in the bone marrow for a long time.

The word *thymus* originated from the Greek word *thymos*, which means soul or spirit. In fact, for centuries, it was believed that the soul was localized in this part of the body. The thymus (figure 5.6) is a bilobed organ found in the space between the sternum and the aorta of the heart. It is within the thymus that progenitor cells are created and then undergo maturation and differentiation into mature T lymphocytes. The thymus is at its largest and most active during the neonatal and preadolescent periods. After this period, the organ gradually disappears and is replaced by fat.

As we age, our immune response capability becomes reduced, which in turn contributes to more infections and a higher risk of cancer (Castelo-Branco and Soveral 2014). The loss of immune function with age is called *immunosenescence*, and as life expectancy across the world continues to increase, so, too, has the incidence of age-related conditions. Emerging studies have observed that this increased incidence correlates with a decrease in T cells, most likely from the thymus atrophying with age and producing fewer T cells to fight off infection (Palmer et al. 2018). The shrinking of the thymus gland begins at birth and is known as *thymic involution*. The rate of thymic T cell production is estimated to decline exponentially over time with a half-life of

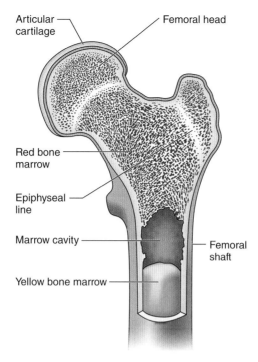

FIGURE 5.5 Bone marrow of the femur.

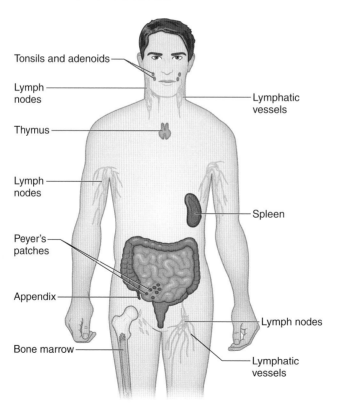

FIGURE 5.6 Primary and secondary lymphoid tissues.

approximately 16 years. It is suggested that the total loss of thymic epithelial tissue and thymocytes would occur at about 120 years of age (Bodey, Siegel, and Kaiser 2006). Thus, this age is a theoretical limit to a healthy human life span.

Secondary Lymphoid Organs

Lymphocytes develop and mature in the primary lymphoid organs, but they mount immune responses from the secondary lymphoid organs, which comprise the spleen, lymph nodes, lymphoid nodules, and lymphoid tissue.

Spleen   The spleen (figure 5.6) is the largest of the lymphatic organs. It is located under the rib cage and above the stomach, to the left of the upper abdomen. In the past it was thought that the spleen was the source of anger, which explains the phrase, *venting your spleen*. While it is an important organ for keeping bodily fluids balanced, it is possible to live without a spleen. The spleen is sometimes called the *filter of the blood* because of its extensive vascularization and the presence of macrophages and dendritic cells that remove microbes and other materials from the blood, including dying red blood cells. Antigen-presenting cells unique to the spleen regulate the T and B cell response to these antigenic targets in the blood. Around 25 percent of our lymphocytes are believed to be stored in the spleen at any one time.

Tonsils   The tonsils (figure 5.6) are lymphoid nodules important in developing immunity to inhaled or ingested pathogens. Technically, there are three sets of tonsils in the body. The palatine tonsils, often simply referred to as the *tonsils*, are situated at the sides of the soft palate. The pharyngeal tonsils, commonly known as *adenoids*, are found on the pharynx behind the nasal cavity. The lingual tonsils are located at the base of the tongue. Each set of tonsils is composed of tissue similar to lymph nodes. Tonsillectomy is the removal of the palatine tonsils and is one of the most common surgical procedures performed on children in the United States. Indications for surgery include recurrent throat infections and obstructive sleep-disordered breathing, both of which can substantially affect a child's health status and quality of life. Controversy persists regarding the benefits of tonsillectomy as compared with observation and medical treatment of throat infections (Mitchell et al. 2019).

Additional Lymphoid Tissue   The appendix is a narrow, blind-ended tube located close to the ileocecal valve that separates the small and large intestine. The function of the appendix has traditionally been a topic of debate. Some of the cells in the mucosa produce organic compounds and hormones to assist with various biological control mechanisms. Its lymphoid tissue is specifically involved with the maturation of B lymphocytes and the production of antibodies (Deshmukh et al. 2014). It is also theorized that the appendix is a safe house for symbiotic gut microbes (Randal Bollinger et al. 2007).

Inflammation of the appendix is known as *appendicitis* and is a common cause for acute, severe abdominal pain. The cause of appendicitis depends on age. In the young, it is mostly due to an increase in lymphoid tissue size, which occludes the lumen.

From 30 years old onwards, it is more likely to be blocked due to hardened feces. The surgical removal of the appendix is called an *appendectomy*. It is now suggested that appendectomies profoundly alter the immune system and modify the pathogenic inflammatory immune responses of the gut (Sanders et al. 2013).

Furthermore, studies have shown that appendectomy-related impairment of the microbes in the gut may lead to microbial imbalance and induce various diseases, including ulcerative colitis, Crohn's disease, *Clostridium difficile* infection, colorectal cancer, rheumatoid arthritis, and cardiovascular disease (Roblin et al. 2012; Sanders et al. 2013; Tzeng et al. 2015; Wu et al. 2015).

Mucosa-associated lymphoid tissue and skin-associated lymphoid tissue mount mucosal and cutaneous responses, respectively, to protect the body tracts and skin. Mucosa-associated lymphoid tissue is present in the bronchi, nasopharynx, and gut.

Can Yoga Help to Reduce Inflammation?

Relationships between mind and body have gradually become more since the early 1990s, and since Ader and Cohen (1975) coined the term *psychoneuroimmunology*, a body of evidence on this interaction has been growing. The field of investigation looking specifically at yoga and immune function is still young, and the current body of evidence is small.

A systematic review of randomized controlled trials by Falkenberg, Eising, and Peters (2018) looked at the relationship between yoga and immune functioning. Fifteen studies met their inclusion criteria, and they concluded that, although the existing evidence is not entirely consistent, a general pattern emerged suggesting that yoga can downregulate proinflammatory markers. They suggested that these results imply that yoga may be implemented as a complementary intervention for populations at risk for or already suffering from diseases with an inflammatory component. The authors hypothesized that longer time spans of yoga practice are required to achieve consistent effects, especially on circulating inflammatory markers.

A meta-analysis by Morgan and colleagues (2014) looked at the effects of mind–body therapies (tai chi, qi gong, meditation, and yoga) on the immune system. Thirty-four studies published in 39 articles (total 2219 participants) met the inclusion criteria. The authors concluded that mind–body therapies, both short term and long term, appear to reduce markers of inflammation and influence virus-specific immune responses to vaccinations. These immunomodulatory effects, albeit incomplete, warrant further methodologically rigorous studies to determine the clinical implications of these findings for inflammatory and infectious disease outcomes.

It is well established that psychological stress and depression impair antiviral immune responses and activate innate immunity or markers of inflammation via effector pathways, such as the sympathetic nervous system and the hypothalamus-pituitary-adrenal axis (Morgan et al. 2014). In chapter 2 on the nervous system, we explore in much more detail the role that yoga can play in combating stress and depression.

CONDITIONS OF THE IMMUNE SYSTEM

We will now explore some of the main conditions that affect the immune system and discuss the potential role that yoga can play here.

Chronic Inflammatory Disease

Chronic inflammatory disease is an overall term for a variety of chronic diseases such as rheumatoid arthritis, asthma, chronic heart failure, chronic obstructive pulmonary disease, cystic fibrosis, type 1 and type 2 diabetes mellitus, inflammatory bowel disease (e.g., Crohn's disease, ulcerative colitis), and multiple sclerosis. Despite common characteristics of systemic inflammation, these disorders have a variety of underlying deficiencies while the precise causes and underlying physiological processes are mostly unknown. In Crohn's disease, multiple sclerosis, rheumatoid arthritis, and type 1 diabetes, chronic systemic inflammation is related to underlying autoimmune disorders whereby the body's immune system attacks its own tissues. Emerging research suggests that stress-related disorders are significantly associated with risk of subsequent autoimmune disease (Song et al. 2018).

Human Immunodeficiency Virus and Acquired Immunodeficiency Syndrome

Human immunodeficiency virus (HIV) attacks the body's immune system, specifically helper T cells, and weakens a person's immunity against infections such as tuberculosis and some cancers. If a person's helper T cell count falls below 200, their immunity is severely compromised, leaving them more susceptible to many infections. Someone with a helper T cell count below 200 or someone who has developed one or more specific opportunistic infections is diagnosed with having acquired immunodeficiency syndrome (AIDS). At the end of 2018, there were 37.9 million people living with HIV (WHO n.d.).

Cancer

The immune system is closely intertwined with both the development of cancer and its treatment. In a way, cancer can be thought of as a manifestation of malfunctions

Can Yoga Specifically Benefit People With HIV?

A systematic review and meta-analysis by Dunne and colleagues (2019) looked at the benefits of yoga for people living with HIV/AIDS. With seven studies meeting the design criteria, the authors concluded that yoga is a promising intervention for stress management. This is particularly significant since stress has been associated with accelerated disease progression for individuals living with HIV (Ironson et al. 2015). However, the literature is limited by the small number of studies, and the authors of the review suggested that randomized controlled trials with objective measures of HIV-related outcomes are needed to further evaluate the benefits of yoga.

The Link Between Insomnia, Immune Function, and Yoga

Lack of sleep has been shown to have substantial adverse consequences for cognitive functioning and metabolic, cardiovascular, immunological, and psychological health (Watson et al. 2017). A systematic review and meta-analysis by Irwin, Olmstead, and Carroll (2016) adds to a growing body of evidence that insomnia is associated with inflammatory disease risk and all-cause mortality, possibly by sleep disturbance affecting inflammatory mechanisms. Interestingly, this study and an additional systematic review and meta-analysis by Cappuccio and colleagues (2010) agree that long sleep duration (greater than eight hours per night) should also be regarded as an additional behavioral risk factor for inflammation.

A systematic review by Wang and colleagues 2015 explored the relationship between tai chi, qi gong, and yoga on sleep quality. The authors reported that the findings of the 17 studies that they included showed that these practices have beneficial effects for various populations on a range of sleep measures. Improvement in sleep quality was reported in the majority of studies and was often accompanied by improvements in quality of life, physical performance, and depression. They concluded that tai chi, qi gong, and yoga may be useful for the treatment of both uncomplicated insomnia and insomnia that was linked to other medical and psychiatric conditions.

In a systematic review looking at yoga and neuropsychiatric disorders, Balasubramaniam, Telles, and Doraiswamy (2013) stated that there is emerging evidence from randomized trials to support popular beliefs about yoga for depression and sleep disorders. In a systematic review and meta-analysis exploring the effect of mind–body therapies including mindfulness meditation and yoga on insomnia, Wang and colleagues (2019) found that the mind–body therapies resulted in statistically significant improvement in sleep quality and reduction in insomnia severity. They concluded that these therapies can be effective in treating insomnia and improving sleep quality for healthy individuals and clinical patients. A systematic review and meta-analysis by Wang and colleagues (2020) demonstrated that yoga intervention in women can be more beneficial than nonactive control conditions in terms of managing sleep problems.

In each of these reviews, the authors noted that studies to date generally have significant methodological limitations. More thorough research is needed to make a better conclusion about the precise effect that yoga can have on insomnia.

---

in immunity, as malignant cells manage to escape recognition and elimination by the immune system. Chronic infections and inflammation associated with limited immune responses can also contribute to the initiation of cancer formation and tumor progression (Shurin 2012). Understanding how the immune system affects cancer development and progression is one of the most challenging questions in immunology. There are also specific cancers that affect the immune system including leukemia and lymphoma. Earlier in the chapter we looked at how yoga can improve the quality of life for people living with cancer.

CONCLUSION

The lymphatic and immune systems are fascinating, complex, and intertwined parts of our physiology. It is reassuring to know that for the vast majority of us, these systems function perfectly day to day. Although some interventions have been found to alter some immune system components, currently there is no evidence that they actually boost immunity to the point where the person is better protected against infection and disease. Yoga is recommended as part of a healthy lifestyle to keep the lymphatic and immune systems working optimally.

Chapter 6

ENDOCRINE SYSTEM

Coming from the words *secrete within,* the endocrine system is one that, when working well, operates largely in the background without us noticing it very much. However, when the endocrine system, which produces hormones that regulate metabolism, sleep, and sexual function, is out of balance, we can become anxious, sleep deprived, and even diabetic. Unlike our thoughts or our breath, we cannot voluntarily control our endocrine system. However, engaging in exercise like hatha yoga can have profound effects on the physiology of this system.

The endocrine system is composed of the pituitary gland, thyroid gland, parathyroid glands, adrenal glands, pancreas, ovaries, and testicles (figure 6.1). These glands produce hormones that regulate metabolism, growth and development, tissue function, sexual function, reproduction, sleep, and mood, among other things. In this chapter, we will look specifically at cortisol, insulin, thyroid hormones, endorphins, and dopamine.

WHAT IS A HORMONE?

Most of us rarely think about hormones except, perhaps, to describe ourselves or someone else in less-than-complimentary terms as *being hormonal.* However, we literally cannot live without hormones. Hormones, appropriately meaning *setting in motion* in Greek, are chemical messengers that are essential to life and dictate how we breathe, how much energy we have available, how we perceive the world, how happy or threatened we feel, and even how we move (Neave 2008; Shuster 2014). While the nervous system is our body's electrical messaging system, hormones are the body's chemical messengers, traveling in the blood to tissues or organs and working more slowly than the electrical impulses of the nervous system. Following a signaling pathway from the brain's hypothalamus to the master gland, the pituitary, hormones are released from various other glands such as the pancreas, the adrenal glands, and the sex glands. Hormones then bind to a variety of receptor tissues, such as muscles, the heart, and the intestines. Hormones are similar to neurotransmitters, and some hormones play both roles. The difference, though, is that hormones are produced in endocrine glands and released into the bloodstream to find their targets of action at

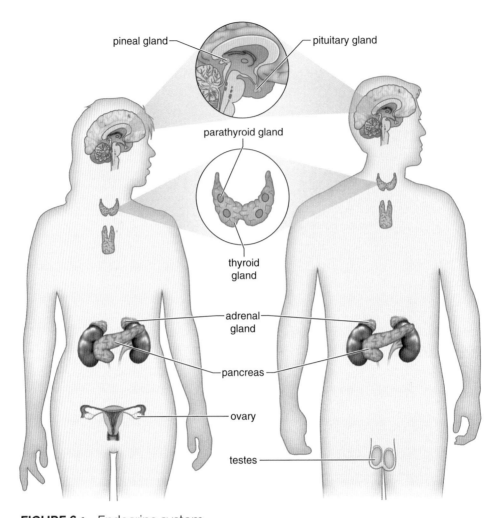

FIGURE 6.1 Endocrine system.

some distance from their origin, whereas neurotransmitters are produced at nerve terminals and released into the local synaptic gap, which is the junction between two neurons.

CORTISOL: THE MASTER HORMONE

Perhaps one of the most famous hormones—often discussed in yoga settings—is cortisol, commonly called the *stress hormone*. While it is true that cortisol levels increase when we are anxious, threatened, or even depressed, cortisol is essential for life. Cortisol is a steroid hormone released from the adrenal glands, which are triangle-shaped organs that sit atop the kidneys, and cortisol literally affects every other system in the body.

The hypothalamus, a small region of the brain, monitors cortisol levels in the blood. If the level is too low, the hypothalamus releases a signal to the pituitary gland, which is immediately below it, and the pituitary gland then releases a hormone signal to the adrenal glands, which alter the amount of cortisol they release (Herman et al. 2011). This three-part system is known as the *hypothalamic-pituitary-adrenal axis* (figure 6.2).

Cortisol receptors, which are in most cells in the body, receive and use the hormone in different ways. Our needs will differ from day to day. For instance, when your body is on high alert, cortisol can alter or shut down functions that get in the way. These might include your digestive or reproductive systems, your immune system, or even your growth processes.

Though cortisol is released in response to stress, the brain also signals the release of cortisol following a circadian (meaning approximately a day) rhythm. Cortisol peaks in the morning, mobilizing our stored energy (fat and glucose) to get us up and

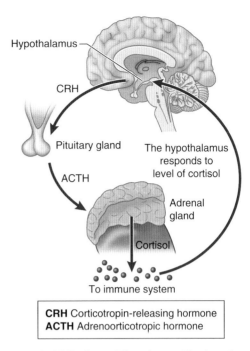

FIGURE 6.2 The hypothalamic-pituitary-adrenal (HPA) axis and the stress response.

moving, and after meals, regulating our blood sugar level, which spikes after we eat.

Cortisol plays other important roles in the body. In addition to helping to control blood sugar levels, which in turn provides the energy necessary for muscular contraction, it also regulates metabolism and blood pressure; it affects sleep quality; it impacts sex drive; it assists with memory making; and it aids in fetal health during pregnancy. Cortisol has anti-inflammatory properties, which can help ease pain and irritation. Interestingly, it is believed that adrenaline and cortisol flood the brain when the news of emotional events is received to create what psychologists call *flashbulb memories*, which are long-term memories noted for their vividness and brevity. For example, you probably remember in great detail where you were when you received the news about the Twin Towers falling in New York City on September 11, 2001, or some other emotionally charged event.

Cortisol is such a powerful substance that it can also be a lifesaving medication. When used as a medication, it is called *hydrocortisone*, and helps to stop a severe allergic reaction like anaphylaxis, which could be a reaction to a peanut or shellfish allergy, for example. It can help treat bouts of rashes and irritation like eczema and psoriasis. Patients with joint pain are sometimes given cortisone injections, which are corticosteroids or hydrocortisone, both closely related to cortisol.

So, if cortisol has so many important and beneficial roles in the body, why does it tend to have a negative connotation? In addition to its everyday functions and circadian release, cortisol is also expressed when we feel threatened. When secreted with a shot of adrenaline, cortisol can help activate our fight-or-flight mechanism to get us out of a dangerous situation—think of a lion running at you or, more appropriate to our modern culture, a bus hurtling toward you. At a time like that, you want your system to respond to get you out of danger as quickly as possible.

Normally considered to be acute hormones, cortisol and adrenaline levels should go down once the threat has passed, and your heart, blood pressure, and other body systems should go back to normal, or into homeostasis. In fact, scientists can measure cortisol levels in blood, saliva, and urine and use these measurements to determine how much stress a subject is under. Long-term heightened cortisol levels can create a number of health problems including anxiety, depression, headaches, heart disease, memory and concentration problems, digestive problems, trouble sleeping, weight gain, and lack of sex drive (American Psychological Association 2018). Some of the effects of long-term elevated cortisol levels are explored in the pathologies section later in this chapter.

INSULIN

Insulin is a hormone made by the pancreas that allows the body to use glucose, or sugar, from the foods we eat or to store that glucose for later use (see chapter 8 on the digestive system for more information about the pancreas). Insulin works to keep the blood sugar from becoming too high (known as *hyperglycemia*) or too low (known as *hypoglycemia*). All cells need glucose for energy, but glucose cannot enter cells directly, so insulin acts as a key that allows the glucose channels to open, letting glucose into the cell (figure 6.3). This release of insulin from the pancreas increases after we eat, especially if we have a meal rich in carbohydrates.

Insulin is quite smart, though, and if we have more sugar in our blood than we currently need, then the same insulin allows the blood sugar to be stored in the liver for later use. So, insulin helps balance out blood sugar levels, keeping them within a normal range.

If the body does not produce enough insulin or the cells are resistant to the effects of insulin, one can develop hyperglycemia (high blood sugar); if

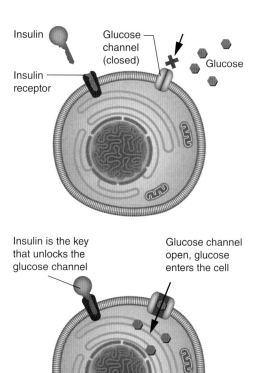

FIGURE 6.3 Insulin acts as a key to allow glucose to enter a cell.

this happens over a long term, complications can arise, including diabetes, which is discussed later in this chapter.

THYROID HORMONES

The thyroid is a gland consisting of two connected lobes that look like butterfly wings at the front of the neck just below your Adam's apple. The thyroid gland secretes three hormones: triiodothyronine (T3), thyroxine (T4), and calcitonin. The first two influence our metabolic rate and protein synthesis as well as growth and development in children, while calcitonin plays a role in calcium homeostasis. As with other glands, the brain's hypothalamus sends a chemical message to the pituitary gland, which then sends a message to the thyroid for the secretion of T3 and T4.

The thyroid hormones play a wide range of roles in our bodies. These include important metabolic roles where the thyroid hormones increase the basal metabolic rate and affect almost all body tissues. They affect appetite, the absorption of substances, and the movements of the gut to digest food as they increase absorption of nutrients in the gut plus the generation of glucose, the uptake of glucose by cells, and the breakdown of glucose. They stimulate the breakdown of fats and increase the number of free fatty acids. Despite increasing free fatty acids, thyroid hormones decrease cholesterol levels.

Thyroid hormones have cardiovascular roles where they increase the rate and strength of the heartbeat. They increase the rate of breathing, the intake and consumption of oxygen, and the activity of mitochondria. Altogether, these factors increase blood flow and the body's temperature.

Thyroid hormones are very important for normal development as they increase the growth rate of young people. Cells of the developing brain are an important target for the thyroid hormones.

Finally, the thyroid hormones also play a role in maintaining normal sexual function, sleep, and thought patterns. Increased thyroid hormone levels are associated with increased speed of thought generation but decreased focus. Sexual function, including libido and the maintenance of a normal menstrual cycle, are also influenced by these essential hormones.

ENDORPHINS: OUR BODY'S MORPHINE

In the early 1970s, researchers were studying how poppy-derived opiates, such as heroin and morphine, affect the brain. They uncovered receptors in the brain that only receive opioids, the class of drugs into which opiates fall. In considering why our brains might have opioid receptors, the researchers hypothesized that our bodies naturally produce substances like morphine and heroin. They were right.

The term *endorphin* is a contraction of *endogenous* (meaning created within the body) and *morphine*. Morphine is used in painkillers like Vicodin (hydrocodone and acetaminophen) and fentanyl, as well as the street drug heroin. It is also chemically similar to the endorphins our bodies produce. They are produced by both the pituitary gland and the central nervous system, so they cannot be easily categorized into just one bodily system.

Functions of Endorphins

Endorphins function as both hormones and neurotransmitters. Researchers are still working to understand all the roles of endorphins. They are considered important in modulating pain and enhancing pleasure, thus promoting an overall sense of well-being. Endorphins play a vital role in pain modulation. Noxious stimuli are received from the peripheral nervous system. The brain decides then whether the experience of pain would be useful at that time. One of the ways the brain will slow down the messages of nociception from the peripheral nervous system is with endorphins transmitted along pathways with serotonin and noradrenaline as neurotransmitters (Akil et al. 1984).

It is believed that endorphins might alleviate depression, reduce anxiety and the stress response, increase self-esteem, and reduce body weight.

Endorphins are important in our natural reward circuits and are related to activities like eating, drinking, physical fitness, and sexual intercourse. They minimize pain and maximize pleasure, which helps us to continue functioning despite injury or stress. Endorphins are also considered responsible for producing the euphoric states experienced with sex, orgasm, listening to music, and eating certain foods like chocolate (Chaudhry and Gossman 2020). Some evidence has also shown that endorphins help reinforce social attachments (Machin and Dunbar 2011). Having strong social bonds certainly helps the survival of a species. Finally, endorphins help with childbirth, which can indeed be a rewarding but painful experience. Pregnant women's endorphin levels rise beyond their normal levels, which may help ease some of the symptoms associated with pregnancy and childbirth (Cahill 1989).

Are Endorphins Responsible for the Runner's High?

Of all the hormones, endorphins are the ones most commonly associated with exercise. The famous runner's high refers to the reduction of anxiety and increased euphoria distance runners have reported in surveys and studies. In the 1980s, exercise scientists started attributing this feeling of blissfulness to endorphins after observing increased levels of these natural painkillers in people's bloodstreams immediately following a run. However, more recent research has revealed that postexercise bliss is more likely due to a different compound: endocannabinoids.

Similar in chemical structure to cannabis, cannabinoids made by our bodies increase in number during pleasurable activities, such as orgasms, and also when we run. First exploring endocannabinoids in mice and more recently in humans, Siebers and colleagues (2021) recruited 63 experienced runners, both male and female, and randomly assigned half to receive naloxone, a drug that blocks the uptake of opioids, which includes endorphins. The other half received a placebo. Most of the subjects reported experiencing a runner's high independent of whether they had taken naloxone or the placebo and all showed increased blood levels of endocannabinoids. This study provides strong evidence that the runner's high previously associated with endorphins is more likely correlated with endocannabinoids, but more research is of course needed to confirm these findings. This study shows that even as recently as 2021, there was still much we did not fully understand about human physiology, and what we previously considered fact may not be so.

Are Endorphins Addictive?

When endorphins latch onto the opioid receptors, they are almost immediately broken down by enzymes, providing the positive benefits of the endorphins and allowing them to be recycled and later reused. However, chemically different but similarly shaped opiates latch onto these same receptors. They are resistant to the enzymes and continually reactivate the receptors, prolonging the high and increasing euphoric feelings as well as the possibility of becoming addicted to that feeling.

The long-term use of artificially produced opioids like morphine and heroin also affects our natural endorphin system. Multiple studies have shown that long-term use of these drugs decrease our body's own production of endorphins. Other studies have also found that the number of opioid receptors of users decreases and that the receptors become less efficient and less sensitive, hence the need of long-term drug users for more drug to get the same high (Sprouse-Blum et al. 2010). Such an outcome is written into our physiology.

Endorphin Imbalances

When out of balance, endorphins may play a role in mental health issues. For example, endorphins may help us to decide when enough is enough, and if someone does not have enough endorphins, as might be the case with obsessive-compulsive disorder, they may never receive the mental cue to stop an activity, such as washing their hands.

Endorphins may also play a role in heightened states of rage or anxiety. If endorphins are overactive or the hypothalamus misreads the endorphin cue, you could be flooded with fight-or-flight hormones with the tiniest of threats. It might be endorphins that trick our brain into feelings of well-being after an inert substance, which we believe to be beneficial, is taken—in other words, the *placebo effect.*

DOPAMINE

Dopamine is another important element of the endocrine system. Dopamine functions as both a hormone and a neurotransmitter and serves many important roles in the brain and body. Dopamine is usually perceived in popular culture as a chemical of pleasure. While it is true that drugs like cocaine cause a quick surge in dopamine levels, dopamine has more to do with motivation than with pleasure. Dopamine seems to be important in signaling whether an outcome is going to be inherently desirable or harmful, which in turn guides our behavior toward or away from that outcome (Wenzel et al. 2015).

Being a neurotransmitter, dopamine is used to send messages between nerve cells, and it plays an important role in our unique human ability to think and plan as it helps us strive, focus, and find things interesting. Dopamine affects many behavioral functions like learning, motivation, mood, and attention, as well as many physical functions like heart rate, blood vessel function, pain processing, and movement, among others. It is most certainly a multifaceted and multifunctional chemical.

TRY IT YOURSELF: Observe Your Thoughts

Knowing that meditation can decrease our cortisol levels and boost dopamine and endorphin levels, here is a basic meditation of thought observation you can try right now and practice regularly.

Sit in a comfortable, upright posture. Close your eyes and take three deep breaths, inhaling through the nose and breathing out slowly through the mouth and making a soft *H* sound as if you were fogging up a mirror. After three breaths like that, let your breath find its own gentle rhythm. Rather than forcing deep breaths, trust that your body knows the best way to breathe. Settle into your breath, observing the gentle rise and fall of the chest and abdomen with each breath. Observe how the air on the nostrils is cooler on the way in and warmer on the way out. Continue observing the breath. As thoughts begin to arise—and they will—simply observe those thoughts without getting enwrapped in them. Think of observing the thoughts as if they were clouds passing in front of you—there one moment and gone the next. Begin to create some distance between yourself and the thoughts, as if you were witnessing them rather than becoming them, and keep returning your awareness to your breath. Remember that it is not wrong to have thoughts, and meditation is not about stopping thoughts. It is instead about noticing your thoughts and choosing where to place your attention. Try this meditation for five minutes, knowing that doing so is providing a host of benefits to your endocrine system.

CONDITIONS OF THE ENDOCRINE SYSTEM

Endocrine disorders can be complex. Here we look at disorders related to cortisol, insulin, thyroid hormones, endorphins, and dopamine, as well as whether yoga can help.

Cortisol and Stress

We are constantly subjected to various stimuli to which our body responds, maintaining homeostasis. You move from a room with less light to one with more light, for example, and the irises of your eyes constrict to adapt to the increased light. As a stimulus becomes more intense, it can incite a stress response in us, releasing stress hormones such as cortisol, adrenaline, and noradrenaline. But the point where a stimulus creates a stressor varies between people, and we have the potential to change our stress response. Furthermore, some degree of stress can be beneficial. Though several definitions exist, *stress* usually refers to the physiological responses that occur when an organism fails to respond appropriately to emotional or physical threats (Selye 1956). It is interesting to note that this definition refers to *stress* as an internal response rather than an external stimulus.

We usually feel stressed when we have too much to do and seemingly not enough time. Also, anticipating a stressful situation can be worse than actually being in one because we can ruminate about it endlessly, secreting stress hormones while doing so. We can all surely think of a time when the period leading up to an event was more stressful than the actual event itself.

The stress response involves activation of the hypothalamic-pituitary-adrenal axis to release cortisol and a cascade of other stress hormones that produce physiologic changes. These hormones can trigger an acute activation of the sympathetic nervous system known as the *fight-or-flight response*. Physiologic changes include an increase in heart rate, breathing rate, blood pressure, blood flow to active skeletal muscles, increased blood sugar to provide more energy to muscles, and increased mental activity, as well as decreased blood flow to the gastrointestinal tract. The overall effect of all these changes is that a person can perform more strenuous activity than normal (Chu, Marwaha, and Ayers 2020).

The stress response is not a bad thing; it is a normal and advantageous reaction to a threat. Exercise of varying intensities also presents various stressors on the body and the stress response can help a person perform better, both in training and in competition. The stress response can even be useful in a dynamic yoga practice as blood is shunted to skeletal muscles. A yoga practice also provides tensile, compressive, and weight-bearing stresses to which our bodies can adapt favorably. However, once the stressor has passed, our stress response should, ideally, calm down, and we return to homeostasis. If a person's stress response does not settle down, though, perhaps because of living with an actual threat such as an abusive home or because of perceiving constant threats from the world, as is often the case with anxiety and posttraumatic stress disorder, serious problems can occur.

Long-term stress that does not ease is associated with a host of problems including irritability, anxiety, depression, headaches, insomnia, digestive problems, weakened immunity, and sexual dysfunction, as well as increased risk of heart attack, stroke, and type 2 diabetes, among others (Chu, Marwaha, and Ayers 2020; Pouwer, Kupper, and Adriannse 2010). Thankfully, there are some things we can do to alleviate a hyperactive stress response.

Research on yoga and stress is still in its infancy, but the small number of quality studies available have shown yoga to be a promising intervention for stress reduction. In 2017, Pascoe, Thompson, and Ski published a meta-analysis of randomized controlled trials comparing yoga asana versus an active control on stress-related physiological measures. Forty-two studies were included in the meta-analysis, and the researchers found that interventions that included yoga asanas were associated with reduced cortisol, blood pressure, resting heart rate, and fasting blood glucose, among other markers associated with the stress response. The authors concluded that practices that include yoga asanas appear to be associated with improved regulation of the sympathetic nervous system and the hypothalamic-pituitary-adrenal system in various populations (Pascoe, Thompson, and Ski 2017).

Even a single session of yoga seems to help with controlling stress. In a 2017 study involving 24 healthy adults by Benvenutti and colleagues, half of the subjects performed a single session of video-instructed hatha yoga while the other half (the control group) watched television. Both groups then performed a stress task involving quick-firing arithmetic in which they were truthfully told that incorrect responses would reduce the amount of money they would receive at the end of the study. The subjects practicing yoga had accelerated blood pressure recovery from stress, reduced salivary cortisol, and increased self-confidence before and after the stress task. This study demonstrated, for the first time, that yoga provides acute, ameliorating effects on the stress response.

Having a sense of purpose can also decrease our stress response. In psychology, eudaemonia refers to well-being derived from self-development, personal growth, and purposeful engagement, qualities often explored in yogic philosophy and practice. In a sample of older women, Ryff, Singer, and Dienberg Love (2004) found that those with higher levels of eudaemonic well-being had lower levels of daily salivary cortisol, inflammatory biomarkers, and cardiovascular risk as well as longer duration REM sleep compared to those showing lower levels of eudaemonic well-being.

We cannot control all the stressful events that come into our lives, but we have the potential to control our response to stress. From Eagle Pose (Garudasana) to Foot-Behind-the-Head Pose (Eka Pada Sirsasana) and everything in between, we often put ourselves in challenging and uncomfortable positions in yoga asana. Then, we steady our mind and our breath to find a sense of calm amid adversity. In this way, yoga can teach us to better understand and control our stress response.

Off the yoga mat, de-stressing can take a variety of forms from having a long bath to playing chess. These activities of self-care, though, are not necessarily self-indulgent, and finding a way to cope with the stressors of life, perhaps through yoga or purposeful living as explored in this chapter, is not a luxury but is essential to long-term health.

Thyroid Disorders

Hyperthyroidism is an overactive thyroid with excessive secretion of thyroid hormones; the most common cause is the autoimmune disorder Graves' disease. This condition often causes a variety of nonspecific symptoms including weight loss, increased appetite, insomnia, decreased tolerance of heat, tremor, palpitations, anxiety, and nervousness. In some cases, it can cause chest pain, diarrhea, hair loss, and muscle weakness.

Hypothyroidism is an underactive thyroid with a deficiency of thyroid hormones. One common cause is iodine deficiency in parts of the world where iodine is lacking. Because iodine deficiency leads to hypothyroidism, iodine is often added to salt (as in, iodized salt) and other foods in certain countries. Hence, hypothyroidism as a result of iodine deficiency is much less common these days. In iodine-sufficient regions, the most common cause of hypothyroidism is the autoimmune disorder Hashimoto's thyroiditis. Autoimmune diseases occur when the body attacks itself, and they can be infamously difficult to diagnose and treat.

An underactive thyroid causes a range of symptoms including fatigue, constipation, dry skin and brittle nails, aches and pains, low mood, cold intolerance, and a slow heart rate. It can be very easy to attribute hypothyroidism symptoms to other health problems.

Hypothyroidism is more common in women. Between the ages of 35 and 65, about 13 percent of women will have an underactive thyroid, and the percentage rises to 20 percent among those over 65 (Harvard Health Publishing 2021). Because the link between hypothyroidism symptoms and thyroid disease is not always obvious, particularly in older people, many women can have an underactive thyroid without knowing it and without the condition being diagnosed or treated.

Shoulder Stand Stimulates the Thyroid Gland

It is commonly said in yoga classes that certain poses affect certain glands, the most common being that Shoulder Stand (Sarvangasana) stimulates the thyroid gland. This connection between Shoulder Stand and the thyroid has been repeated so many times that many consider it fact, which constitutes a form of cognitive bias known as *availability cascade*, wherein a collective belief gains more and more acceptance simply through its increasing repetition in public discourse. However, the claim of Shoulder Stand stimulating the thyroid gland has not been investigated scientifically and is based purely on speculation (Pierce 2011). While some studies have been conducted on thyroid conditions and yoga in general, we authors cannot find any studies in the scientific literature looking specifically at how Shoulder Stand might affect the thyroid gland. Furthermore, the idea that this pose might stimulate the thyroid gland does not make a lot of sense physiologically.

The workings of the endocrine system are much more complex than is suggested with this idea of applying manual pressure to create change. The endocrine system works through molecular and cellular processes in which one molecule initiates a cascade of events to create the desired outcome. As laid out in this chapter, the pituitary is the master that governs all other glands, and the pituitary is governed by the hippocampus in the brain. The hippocampus releases thyrotropin-releasing hormone, which is received by the pituitary, which then releases thyroid-stimulating hormone, which is then received by the thyroid to produce thyroid hormones. If the hippocampus' ability to produce thyrotropin-releasing hormone were impaired or the pituitary's ability to produce thyroid-stimulating hormone were impaired, as might be the case if a tumor were present, the thyroid would not be able to produce the needed thyroid hormones no matter how many Shoulder Stands were performed. Similarly, if one's diet were deficient in iodine, one's thyroid would not be able to produce enough thyroid hormones, no matter what asanas were performed.

The reality is that there is no scientific evidence to support the notion that Shoulder Stand might directly affect thyroid function. Furthermore, does the thyroid need stimulating? Wouldn't that depend on whether someone has an underactive versus overactive thyroid? On the other hand, just because there is no scientific evidence to support a claim does not mean that the claim is false. What we do know is that all moderate exercise will have some positive effect on the endocrine system, including the thyroid. General exercise has near-miraculous health benefits, even though it does involve applying pressure to the thyroid. We also know that elevated cortisol negatively impacts the thyroid, so taking time to relax, as we do in yoga, also benefits thyroid function. Simply moving and balancing that movement with relaxation are the best things we can do for overall health and among the best gifts we can offer to others.

A very small number of studies exist on the effects of yoga on thyroid conditions. Singh and colleagues (2011) looked at the effect of yoga on the quality of life of 20 female patients with hypothyroidism. After attending one-hour yoga sessions daily for one month, patients' quality of life scores improved, and they reported significant improvement in their perception of their health. The authors conclude that yoga is valuable in helping hypothyroid patients manage their disease-related symptoms, and it may be considered as a supportive therapy in conjunction with medical therapy for the treatment of hypothyroidism (Singh et al. 2011). This was a small study and lacked a control group, but it offers some promise that yoga might be beneficial in the quality of life of people with thyroid disorders.

Chronic stress also makes thyroid conditions worse, and yoga is known to help people deal with long-term stress (Pascoe, Thompson, and Ski 2017). One of the main ways yoga might improve thyroid health is through helping people handle stress.

While there is scant research about thyroid conditions and yoga, a considerable amount of literature shows the benefits of exercise in general on thyroid conditions. A study in 2015 compared the effects of regular exercise on people being treated for underactive thyroid (Bansal et al. 2015). They found that the thyroid hormones T3 and T4 were significantly raised in the exercising group and not in the nonexercising group, concluding that exercise can be a useful treatment alongside the taking of medication.

Altaye and colleagues (2019) explored the effects of exercise on thyroid concentrations of adolescents with intellectual disabilities. They found that after 16 weeks of exercise, a more significant change was observed in the plasma level concentration of thyroid hormones (T3 and T4) of the subjects doing the exercise in comparison to the control subjects, concluding that aerobic exercise had a positive impact on the thyroid hormones.

When looking at exercise intensity and thyroid levels, Ciloglu and colleagues (2006) found that exercise improves thyroid hormones and that 70 percent of maximal effort seems to provide the most results. After that are diminishing returns. In addition to the evidence that exercise directly affects thyroid health, we also know that exercise's secondary benefits certainly help people with thyroid conditions, and these benefits include boosting mood, helping to lose weight, and increasing energy.

Diabetes Mellitus

Diabetes mellitus, commonly known as *diabetes*, is a disease characterized by long-term high blood sugar, and it is becoming dangerously common. A metabolic disease of the endocrine system, diabetes is due to either the pancreas not producing enough insulin or the target cells of the body not responding properly to the insulin produced. Symptoms often include frequent urination, increased thirst, and increased appetite. Long-term complications from diabetes include cardiovascular disease, stroke, kidney disease, foot ulcers, damage to the nerves, damage to the eyes, and cognitive impairment. An acute complication, sometimes called a *diabetic attack*, can include vomiting, abdominal pain, deep gasping, confusion, and, occasionally, loss of consciousness.

Diabetes presents in three different forms. With type 1 diabetes, which was once known as *juvenile diabetes* or *insulin-dependent diabetes*, the pancreas produces little or no insulin by itself. The more common type 2 diabetes, formerly called *adult-onset diabetes* until children began presenting with it, occurs when the body becomes resistant to insulin or does not make enough insulin. Type 2 diabetes has risen dramatically in the last 30 years in countries of all income levels (Shaw, Sicree, and Zimmet 2010) and is believed to be connected to sedentary lifestyle and poor diet. However, we also know that increased cortisol levels create insulin resistance, and so obesity and diabetes might be a response to a stressful life. Finally, gestational diabetes is the third main form. It occurs when pregnant women without a previous history of diabetes develop high blood sugar levels but lower than the standard diabetic levels.

Diabetes is a global problem. As of 2019, an estimated 463 million people had diabetes worldwide, which is a substantial 8.8 percent of the adult population, with type 2 diabetes making up about 90 percent of the cases. Current trends suggest that these rates will continue to rise (International Diabetes Federation 2019). With it at least doubling a person's risk of early death and being the seventh leading cause of death globally, diabetes certainly is a contemporary problem. But exercise, including yoga, can help profoundly.

Exercise and yoga can help to increase the sensitivity of the tissues' receptor cells to the hormone insulin, which in turn helps to control blood glucose (sugar) levels and even increase the total number of receptor cells on the target tissue (Bird and Hawley 2017). Can yoga specifically help? While conventional exercise is known to be beneficial for people with diabetes, there has been much less research about yoga and diabetes, but some studies have shown promise. In 2008, Gordon and colleagues compared the effects of physical activity on patients with diabetes. The 77 participants were divided into three groups as follows:

1. Participants in one group attended a weekly two-hour yoga session and were encouraged to practice yoga three or four times per week at home.

2. Participants attended a two-hour session of conventional physical training including aerobic walking, aerobic dance, and flexibility exercises, and they were encouraged to exercise at a similar rate three or four times per week at home.

3. The control group followed a treatment plan as recommended by their physicians but were not engaged in any kind of active exercise intervention.

The results were very positive.

Knowing that diabetes is equated with elevated blood glucose, seeing a reduction in fasting blood glucose levels is a very promising finding. In this study, the yoga participants had a 29.48 percent reduction and the conventional exercise group saw a 27.43 percent reduction compared to a 7.48 percent reduction in the control group. Both the yoga and the conventional exercise groups saw a reduction in total cholesterol, while the control group saw an increase. Similarly, though the changes were not enough to be considered significant, both exercise groups saw a decrease in their blood triglycerides while the control group saw an increase. Other beneficial results were recorded from the two exercise groups. This study shows that yoga can be as beneficial as conventional exercise at improving the biomarkers that can lead to diabetic complications and early death.

Even more important than any one study, however, is a review and meta-analysis of several studies, and in 2017, Cui and colleagues did just that. These researchers, based in China, looked at a total of 12 randomized controlled trials with a total of 864 patients to evaluate the efficacy of yoga in adults with type 2 diabetes mellitus. They found that, based on the evidence, yoga significantly reduces fasting blood glucose levels and alters other significant clinical outcomes in patients with type 2 diabetes mellitus. These data support the idea that yoga-based training is a viable alternative exercise for type 2 diabetes management. However, given the minimal amounts of research on yoga and type 2 diabetes, they also recommended that more large-scale and robust randomized controlled trials must be conducted.

If you search the Internet for yoga and diabetes, you will find results with videos like "These 5 Yoga Poses Will Cure Your Diabetes" or "Do This 1 Exercise to Help Diabetes." It is easy to think that certain yoga poses might help with certain conditions like diabetes, but the reality is any physical activity is going to have a positive effect.

Dopamine Imbalances

While mental health disorders are due to many causes, they are often linked to deficient or excessive dopamine in different parts of the brain. Some cases of schizophrenia are due to an excess of dopamine in certain parts of the brain and can lead to hallucinations and delusions, while a lack of it in other parts can cause different symptoms, such as lack of motivations and desire (Ayano 2016). Dopamine levels decline by around 10 percent per decade from early adulthood and have been associated with declines in cognitive and motor performance (Peters 2006).

No one knows for sure the causes of attention-deficit/hyperactivity disorder, but some research shows it may be due to a shortage of dopamine. Methylphenidate (Ritalin), an attention-deficit/hyperactivity disorder drug, works by boosting dopamine. Drugs like cocaine cause a quick surge of dopamine in the brain, which can satisfy the natural reward system. The amount of dopamine connecting to receptors in the brain after a dose of cocaine can exceed the amounts associated with natural activities, producing pleasure greater than that which follows thirst-quenching or sex. In fact, some laboratory animals, if given a choice, will ignore food and keep taking cocaine until they starve. But repeated drug use also raises the threshold to get the same high, so users need to take more to get the same feeling. Meanwhile, drugs make your body less able to produce dopamine naturally, which can lead to emotional lows and long-term neurological consequences (Enevoldson 2004).

Imaging studies suggest that people with obesity could have problems with their natural reward systems so that their body may not release enough dopamine and serotonin. If that is the case, then the simple advice to just eat less does not account for the complexities of obesity (van Galen et al. 2018).

Parkinson's Disease

Parkinson's disease (PD), a condition put in the spotlight by actor Michael J. Fox, who was diagnosed with it in 1991 at just 29 years old, is a long-term, degenerative disorder involving tremors and motor impairment. It is caused by a loss of dopamine-secreting neurons in an area of the midbrain called the *substantia nigra*. Dopamine enables neurons in the brain to communicate and control movement, and with PD, the brain makes less dopamine. This chemical deficiency causes physical symptoms such as tremor, stiffness, slowness of spontaneous movement, poor balance, and poor coordination. However, along with medication, exercise, yoga, and meditation can provide some hope for slowing the progression of the disease.

Several studies have shown that exercise can improve quality of life for people with PD. Balance training has been shown, unsurprisingly, to improve balance in patients with PD and, importantly, reduce their fall rates, which is a worthwhile endeavor as falls are quite common among people with PD and can lead to major problems. Sustained tai chi, dance, and resistance training have also been seen to alleviate PD motor symptoms, which suggests that such forms of exercise could slow down the progression of PD (Mak and Wong-Yu 2019). Various other studies have shown that regular intense exercise several times per week significantly improves motor control in people with Parkinson's disease, which suggests that there may be a beneficial effect on the dopamine system (Fisher et al. 2013; Petzinger et al. 2015).

Yoga seems to be able to help with the regulation of dopamine levels. Pal and colleagues (2014) found that one hour of yoga six days a week significantly increased dopamine levels. Similarly, another study on eight experienced meditation teachers found they had a 65 percent increase in dopamine production after meditating for one hour (Kjaer et al. 2002). However, it is unclear whether novice practitioners of meditation can expect similar results. What is clear is that in exercising, practicing yoga, or meditating, a lot more is going on than just moving muscles and bones or shutting the eyes and relaxing, and there is still much more to research.

Depression and Other Mental Disorders

The vast majority of research on the topic of depression shows that exercise has beneficial effects on people with depression, and some researchers hypothesize that this may, at least in part, be down to how exercise affects the endocrine system. Brinsley and colleagues (2020) conducted a systematic review and meta-analysis on the effects of yoga on depressive symptoms in people with mental disorders. Nineteen studies were included in the review, and 13 studies met the inclusion for the meta-analysis, looking at disorders of depression, posttraumatic stress, schizophrenia, anxiety, alcohol dependence, and bipolar disorder. Yoga showed greater reductions in depressive symptoms than treatment as usual, wait-listing control (putting patients on a wait list but not giving a treatment), and attention control (including health education and social support). Furthermore, the researchers found that the more frequently the people participated in yoga, the greater the reduction in their depressive symptoms (Brinsley et al. 2020).

We also know that people with depression, as well as those with other mental illnesses, do less exercise in general than healthy people. Vancampfort and colleagues (2017) found in a review and meta-analysis of the literature that people with schizophrenia, bipolar disorder, and major depressive disorder spend an average of nearly 8 hours of their waking day being sedentary and spend a mean of just 38.4 minutes per day in moderate or vigorous physical activity, which makes them significantly more sedentary and significantly less active than their healthy counterparts. The evidence supports the idea that exercise, including yoga, is an important element in the treatment of various mental disorders.

CONCLUSION

With cocaine affecting dopamine, heroin and other opioids mimicking endorphins, and marijuana mimicking endocannabinoids, it is clear that our brains have their own medicine cabinets providing feelings of reward, euphoria, and calm. Many of these hormones, including the stress hormone cortisol, are powerful substances that are important for overall health, and our body normally does a good job of maintaining these chemicals in delicate balance. More than 50 known hormones circulate in our blood, and in this chapter, we have looked individually at just a few, examining the physiology of each as well as how imbalances can have deleterious effects and even lead to mental illness. However, it is important to remember that all these hormones function together in the complex machinery of the endocrine system. While our brains control the regulation of the endocrine system, what we do with our bodies matters. Though there is less research available on how exercise affects the endocrine system compared to the research on other bodily systems, it is probably safe to say that exercise including yoga is very beneficial to this system. Exercise can help balance our internally produced chemicals. Yoga seems to be effective in treating some endocrine-related conditions, including depression.

The endocrine system seems an important player in our overall happiness and well-being. It is understood that genetics play a major role in our happiness, but biological and health factors are critical in underlying happiness (Dfarhud, Malmir, and Khanahmadi 2014). Happiness cannot be traced to one gene or to one hormone but is affected by many hormones and neurotransmitters like cortisol, dopamine, serotonin, endorphins, and endocannabinoids working together and playing different roles. While we cannot change the genes we have inherited, we can positively affect the balance of our hormones by maintaining a healthy lifestyle that includes plenty of physical activity.

Chapter 7

REPRODUCTIVE SYSTEM

Many practitioners begin their yoga journey during pregnancy as a way to maintain a healthy mind and body and with the hopes of feeling more relaxed and prepared for birth. In this chapter, we will look at the benefits that yoga can offer an expectant mother and explore many other topics including whether it is okay to invert the body during menstruation, whether yoga can help in the management of the common symptoms experienced during menopause, and whether yoga can help couples to manage the stress of fertility treatments.

FEMALE REPRODUCTIVE SYSTEM ANATOMY

The female reproductive system (figure 7.1) functions to produce oocytes (pronounced "oh-uh-sites"), which are immature ova, or egg cells, and reproductive hormones. It has the essential task of supporting the developing fetus and delivering it to the outside world. Unlike its male counterpart, the female reproductive system is located primarily inside the pelvic cavity. We will now look at the main organs of the reproductive system in greater detail.

Vulva

The external female genitalia are collectively called the *vulva*, which is made up of two folds of skin: the labia minora and labia majora. The labia minora serve to protect the female urethra and the entrance to the female reproductive tract. The superior, anterior portions of the labia minora come together to encircle the clitoris (or glans clitoris), an organ that originates from the same cells as

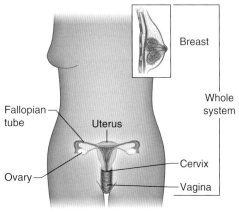

FIGURE 7.1 Female reproductive system.

the glans penis and has abundant nerves that make it important in sexual sensation and orgasm. The labia majora cover and protect the inner, more delicate, and sensitive structures.

Vagina

The vagina is a muscular canal that serves as the pathway into and out of the uterus. The walls of the vagina have smooth muscle that allows for expansion to accommodate intercourse and childbirth. The vagina is home to a normal population of microorganisms that help to protect against infection by pathogenic bacteria, yeast, or other organisms that can enter the vagina. The superior portion of the vagina, called the *fornix*, meets the protruding uterine cervix.

Ovaries

The paired ovaries are each about the size of an almond and produce oocytes. The grouping of an oocyte and its supporting cells is called an *ovarian follicle*. Later in the chapter, we will explore the process of the ovarian cycle in detail. The ovulated oocyte with its surrounding cells is picked up by the funnel-shaped cavity of the uterine tube, and beating cilia (hair-like structures) help to transport it through the tube toward the uterus. Following ovulation, the cells of the empty follicle transform into the progesterone-producing corpus luteum.

Uterine Tubes

The uterine tubes are commonly referred to as the *fallopian tubes* and serve as the passageways that carry the oocyte from the ovary to the uterus. High concentrations of estrogen around the time of ovulation induce contractions of the smooth muscle along the length of the uterine tube that result in a coordinated movement. This movement sweeps the surface of the ovary and the pelvic cavity, and, along with the coordinated beating of cilia that line the outside and lumen of the uterine tube, the oocyte is pulled into the interior of the tube. When fertilization does occur, sperm typically meet the oocyte while it is still moving through the middle section of the uterine tube. If the oocyte is successfully fertilized by a sperm, the resulting zygote will begin to divide as it makes its way through the remainder of the uterine tube and into the uterus, where it will implant and continue to grow.

Uterus

The uterus is the muscular organ that nourishes and supports the growing embryo. It is mostly made up of a thick layer of smooth muscle along with the endometrium—the innermost layer. As the embryo moves through the uterine tubes, the endometrium proliferates, changes in shape, becomes receptive to implantation, and produces a hospitable environment for the embryo. There is only a brief window between 6 and 10 days after ovulation in which the zygote can implant on the endometrium. If an embryo implants into the wall of the uterus, signals are sent to the corpus luteum to continue secreting progesterone to maintain the endometrium, and thus maintain the pregnancy.

The cervix is the narrow inferior portion of the uterus that projects into the vagina. The cervix produces mucus secretions that become thin and stringy under the influence of high levels of estrogen, and these secretions can facilitate sperm movement through the reproductive tract.

Breasts

The breasts are considered accessory organs of the female reproductive system because of their important role in supplying milk to an infant in a process called *lactation*. Breast milk is produced by the mammary glands within the breast tissue, which are modified sweat glands. During the normal hormonal fluctuations in the menstrual cycle, breast tissue responds to changing levels of estrogen and progesterone. This can lead to swelling and breast tenderness in some individuals. If pregnancy occurs, the increase in hormones leads to further development of the mammary tissue and enlargement of the breasts.

There are many documented short- and long-term medical and neurodevelopmental advantages of breastfeeding that are summarized in a paper by the American Academy of Pediatrics (2012). The paper states that infants who are breastfed have reduced risks of asthma, obesity, type 1 diabetes, severe lower respiratory disease, acute otitis media (ear infections), sudden infant death syndrome, and gastrointestinal infections (diarrhea and vomiting). Higher intelligence scores are also noted in infants who are exclusively breastfed for three months or longer. Breastfeeding also appears beneficial for the mother as it can help lower her risk of high blood pressure, type 2 diabetes, ovarian cancer, and breast cancer.

The American Academy of Pediatrics recommends exclusive breastfeeding for about six months, followed by continued breastfeeding as complementary foods are introduced, with continuation of breastfeeding for one year or longer as mutually desired by mother and infant. It is important to add here that breastfeeding is often a complex and challenging process. Even if a mother cannot breastfeed for the full six months to one year, anything is better than nothing, and the first six weeks are particularly valuable. For mothers who struggle with breastfeeding, help is often available, and some countries even have breastfeeding help lines. Yoga classes for parents and their babies can also be a great support.

FEMALE REPRODUCTIVE SYSTEM PHYSIOLOGY

We will now explore the physiological processes of the female reproductive system including the ovarian cycle, menstruation, pregnancy, and menopause.

Ovarian Cycle

The ovarian cycle is a set of predictable changes in a female's oocytes and ovarian follicles. During a woman's reproductive years, the ovarian cycle is roughly 28 days. It includes two interrelated processes: oogenesis (the production of oocytes) and folliculogenesis (the growth and development of ovarian follicles) (figure 7.2).

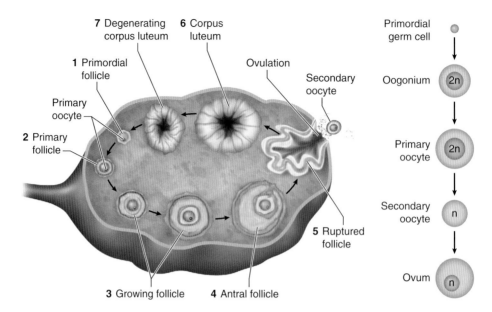

FIGURE 7.2   Oogenesis and folliculogenesis.

Oogenesis

Primary oocytes are present even before birth, but their development is arrested until puberty. The number of primary oocytes present in the ovaries declines from 1 to 2 million in an infant, to approximately 400,000 at puberty, to zero by the end of menopause. During puberty and throughout a woman's reproductive years, surges of luteinizing hormone (LH) initiate the transition from primary to secondary oocyte. The subsequent release of an oocyte from the ovary is called *ovulation*; it occurs approximately once every 28 days. The initiation of ovulation marks the transition from puberty into reproductive maturity for women.

Folliculogenesis

Follicles in a resting state, known as *primordial follicles*, are present in newborn females. They have only a single flat layer of support cells that surround the oocyte. After puberty, each day, a few primordial follicles begin to develop into primary follicles, then secondary follicles and eventually tertiary, or antral, follicles. Early tertiary follicles are stimulated to grow by an increase in follicle-stimulating hormone (FSH) produced by the anterior pituitary gland, and supporting cells in the growing follicles are stimulated by LH to produce estradiol, a type of estrogen.

When the level of estrogen in the bloodstream is high enough, it triggers the pituitary gland (via the hypothalamus) to reduce the production of LH and FSH, and as a result of this, most tertiary follicles in the ovary die. One follicle, usually the one with the most FSH receptors, survives this period and is called the *dominant follicle*. The dominant follicle produces so much estrogen that the pituitary gland is triggered to release large amounts of LH and FSH. The LH surge induces ovulation.

Following ovulation, the cells of the empty follicle transform into the progesterone-producing corpus luteum. Progesterone is a hormone that is critical for the establishment and maintenance of pregnancy.

Menstruation

If an embryo does not implant into the uterine lining, no signal is sent to the corpus luteum, and it degrades, ceasing progesterone production and ending that phase of the ovulation cycle. Without progesterone, the endometrium thins, and the spiral arteries of the endometrium constrict and rupture, preventing oxygenated blood from reaching the endometrial tissue. As a result, endometrial tissue dies, and blood, pieces of the endometrial tissue, and white blood cells are shed through the vagina during menstruation.

The first commencement of menstruation is referred to as *menarche*. A study by Chumlea and colleagues (2003) reported that fewer than 10 percent of girls in the United States start to menstruate before 11 years, and 90 percent of all girls in the United States are menstruating by 13.75 years of age, with a median age of 12.43 years. This age of menarche is not significantly different than that reported for U.S. girls in 1973, despite some concern that puberty is happening earlier now for many girls.

Menstrual Cycle

The timing of the menstrual cycle (figure 7.3) starts with the first day of bleeding, referred to as *day one* of the cycle. Cycle length is determined by counting the days between the onset of bleeding in two subsequent cycles. Typically, the average length of a woman's menstrual cycle is 28 days, and this is the time period used to identify the timing of events in the cycle. However, the length of the menstrual cycle varies among women, and even in the same woman from one cycle to the next, typically from 21 to 32 days.

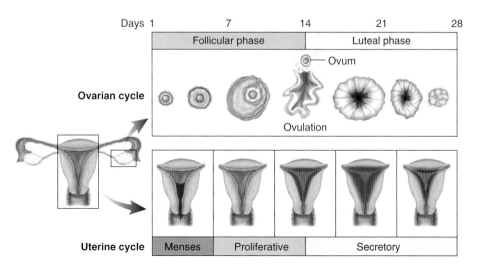

FIGURE 7.3 Menstrual cycle.

Just as the hormones produced by the cells of the ovary trigger the follicular and luteal phases of the ovarian cycle, they also control the three distinct phases of the menstrual cycle: the menses phase, the proliferative phase, and the secretory phase. The menses phase of the menstrual cycle is the phase during which the lining is shed; that is, the days that the woman menstruates. Although it averages approximately five days, the menses phase can last from two to seven days, or longer. Once menstrual flow ceases, the endometrium begins to proliferate again, marking the beginning of the proliferative phase of the menstrual cycle. In the uterus, progesterone from the corpus luteum begins the secretory phase of the menstrual cycle, in which the endometrial lining prepares for possible implantation.

Does the Lunar Cycle Affect the Menstrual Cycle and Our Basic Physiology?

The terms *menstruation* and *menses* come from Latin and Greek words meaning month (*mensis*) and moon (*mene*). While the lunar cycle and the average menstrual cycle are basically equal in length, research has found that the lunar phase does not influence menstruation (Binkley 1992). The app Clue allows women to track their periods and ovulation cycles. Their data science team analyzed 7.5 million menstrual cycles and found no correlation between the lunar phases and the menstrual cycle or period start date (Clue 2019).

It has always been the tradition in ashtanga yoga to rest from asana practice on new- and full-moon days. Tim Miller has been studying and teaching ashtanga yoga since 1979 and was the first American certified to teach by the Ashtanga Yoga Research Institute in Mysore, India. This is how Miller explains the tradition of moon days:

> *The phases of the moon are determined by the moon's relative position to the sun. Full moons occur when they are in opposition and new moons when they are in conjunction. Both sun and moon exert a gravitational pull on the earth. Their relative positions create different energetic experiences that can be compared to the breath cycle. The full moon energy corresponds to the end of inhalation when the force of prana is greatest. This is an expansive, upward moving force that makes us feel energetic and emotional, but not well grounded. During the full moon we tend to be more headstrong. The new moon energy corresponds to the end of exhalation when the force of apana is greatest. Apana is a contracting, downward moving force that makes us feel calm and grounded, but dense and disinclined toward physical exertion. Observing moon days is one way to recognize and honor the rhythms of nature so we can live in greater harmony with it. (Ashtanga Yoga Center, n.d., para. 2-4)*

Despite a popular belief that our mental health and other behaviors are modulated by the phase of the moon, there is no solid evidence that human biology is in any way regulated by the lunar cycle. Although the moon clearly influences oceanic tides, it does not produce tides in smaller bodies of water such as lakes and even some seas, let alone in a human body (Culver, Rotton and Kelly 1988). Foster and Roenneberg (2008) state that the gravitational forces that generate the tides depend on the distance between Earth and the moon; on the alignment of the moon, Earth, and the sun; but not on the phases of the moon. Therefore, a full moon does not have a specific gravitational

Inversions Cause Reversal of Blood Flow During Menstruation

There is widespread belief among yoga teachers and practitioners that women should not practice inversions including Shoulder Stand (Salamba Sarvangasana) and Headstand (Sirsasana) during their period. The reasoning behind this often has to do with the perceived flow of energy throughout the body. On an energetic level, menstruation is intrinsically linked with apana (the contracting, downward moving force), and therefore it is believed that inverting the body will reverse the direction of this force. Another part of the reasoning comes from the retrograde menstruation theory, which was promoted by Doctor John Sampson (1927). In 1927, Sampson suggested that menstrual tissue can flow backward through the fallopian tubes and deposit on the pelvic organs, causing endometriosis (the abnormal growth of endometrial cells outside the uterus). However, there is little evidence that endometrial cells behave in this way. More recently, researchers have found that up to 90 percent of women have retrograde flow anyway (Sasson and Taylor 2008), but since only 10 percent of women of reproductive age develop endometriosis (Olive and Schwartz 1993), it has been concluded that the cause of endometriosis is much more complicated than this. Sampson's theory has also been disputed, because it cannot explain the occasional occurrence of endometriosis in prepubertal girls, newborns, women who have had a hysterectomy (surgical removal of the uterus), and men who have received long-term hormonal treatment. Writing about female astronauts in space, Wotring (2012) reported that the myth that zero gravity would cause retrograde menstrual flow, causing blood to accumulate in the abdomen and cause infections, has been shown to be baseless. It is understood that uterine contractions, rather than one's orientation to the ground, are responsible for the flow of menstrual blood (Bulletti et al. 2000).

It is also important to recognize that the uterus is inverted many times throughout a typical yoga practice—for example, in Standing Forward Fold (Uttanasana), Downward Facing Dog (Adho Mukha Svanasana), Bridge Pose (Setu Bandha Sarvangasana), and others—yet these poses are rarely classed as inversions. Thus, there is little logic to the idea that inverting the uterus in these poses is appropriate but not appropriate during a Shoulder Stand. There is rarely, if ever, a single directive that works for every student. It is important for yoga practitioners to tune into to what feels right for them in each moment and for yoga teachers to give options and permission for students to do so.

---

effect on Earth. A literature review by the same authors also confirmed that the lunar phase has no effect on conception (in vitro fertilization), the number of births occurring, psychosis, depression, anxiety, violent behavior or aggression, seizures, suicide, coronary failure, or automobile accidents. However, a study by Cajochen and colleagues (2013) suggested that the lunar cycle modulates human sleep and melatonin rhythms. They found that around the full moon, deep sleep levels decreased by 30 percent, time to fall asleep increased by five minutes, and total sleep duration was reduced by 20 minutes. These changes were associated with a decrease in subjective sleep quality and diminished endogenous melatonin levels. This is the first reliable evidence that a lunar rhythm can modulate sleep structure in humans when measured under the highly controlled conditions of a circadian laboratory study protocol without time cues.

Pregnancy

A full-term pregnancy lasts approximately 38.5 weeks from conception to birth, but since it is easier to pinpoint the first day of the last menstrual period, the expectancy date is normally set as approximately 40.5 weeks from the last menstrual period. This assumes that conception occurred on day 14 of the woman's cycle. The 40 weeks of an average pregnancy are usually described in terms of three trimesters (figure 7.4), each approximately 13 weeks.

As the placenta develops, it gradually takes over from the degenerating corpus luteum as the endocrine organ of pregnancy. Estrogen maintains the pregnancy, promotes fetal viability, and stimulates tissue growth in the mother and developing fetus. Progesterone prevents new ovarian follicles from developing and suppresses uterine contractility until labor. The hormone relaxin has beneficial effects on the endometrium responsible for establishment of pregnancy. This hormone also stimulates endometrial decidualization—a process that results in significant changes to cells of the endometrium in preparation for and during pregnancy. This includes morphological and functional changes to endometrial cells, the presence of white blood cells, and vascular changes to maternal arteries (Goldsmith and Weiss 2009).

The second and third trimesters of pregnancy are associated with dramatic changes in maternal anatomy and physiology. The most obvious anatomical sign of pregnancy is of course the enlargement of the abdominal region, coupled with maternal weight gain. This weight results from the growing fetus as well as the enlarged uterus, amniotic fluid, and placenta. Additional breast tissue and dramatically increased blood volume also occur.

Nausea and vomiting in pregnancy typically begin between the fourth and seventh week after the last menstrual period in 80 percent of pregnant women and resolves by the 20th week of gestation in all but 10 percent of these women (Gadsby, Barnie-Adshead, and Jagger 1993). The cause of nausea and vomiting during pregnancy remains unknown, but a number of possible causes have been investigated including decreased intestinal peristalsis, increased circulation of pregnancy-related hormones, and chronic gut bacterial infection (Quinlan and Ashley Hill 2003). Other common side effects during pregnancy include gastric reflux, or heartburn, which results from the upward, constrictive pressure of the growing uterus on the stomach; constipation from decreased intestinal peristalsis; more frequent urination due to the compression of the bladder by the uterus; and an increase in the total amount of urine produced.

Blood volume can increase by up to 30 percent, which also leads to an increase in heart rate and blood pressure. Venous return is often impacted, and varicose veins or hemorrhoids can develop. The pressure of the growing fetus on the diaphragm can also limit the volume of inhalation and cause shortness of breath.

It is also important to highlight that there are many possible benefits of being pregnant. Many women report feeling very energized and positive in their outlook, particularly during the second and third trimesters. Self-confidence can improve, senses can become heightened, and many women find being pregnant an absolute joy.

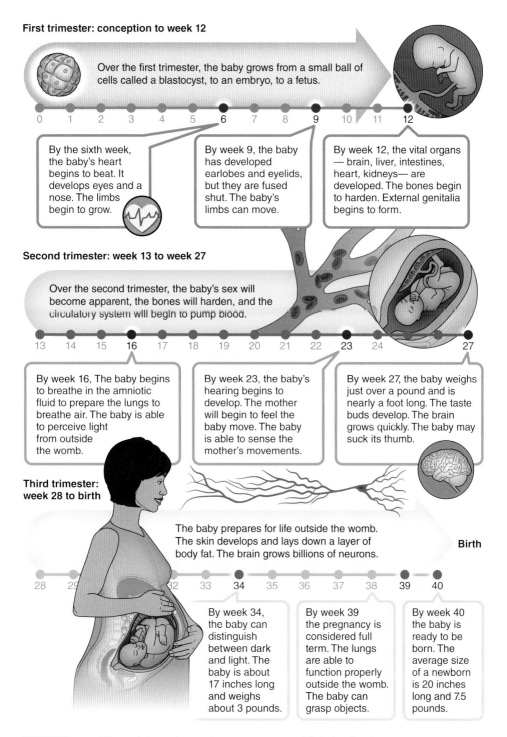

First trimester: conception to week 12

Over the first trimester, the baby grows from a small ball of cells called a blastocyst, to an embryo, to a fetus.

0   1   2   3   4   5   6   7   8   9   10   11   12

By the sixth week, the baby's heart begins to beat. It develops eyes and a nose. The limbs begin to grow.

By week 9, the baby has developed earlobes and eyelids, but they are fused shut. The baby's limbs can move.

By week 12, the vital organs — brain, liver, intestines, heart, kidneys— are developed. The bones begin to harden. External genitalia begins to form.

Second trimester: week 13 to week 27

Over the second trimester, the baby's sex will become apparent, the bones will harden, and the circulatory system will begin to pump blood.

13   14   15   16   17   18   19   20   21   22   23   24   27

By week 16, The baby begins to breathe in the amniotic fluid to prepare the lungs to breathe air. The baby is able to perceive light from outside the womb.

By week 23, the baby's hearing begins to develop. The mother will begin to feel the baby move. The baby is able to sense the mother's movements.

By week 27, the baby weighs just over a pound and is nearly a foot long. The taste buds develop. The brain grows quickly. The baby may suck its thumb.

Third trimester: week 28 to birth

The baby prepares for life outside the womb. The skin develops and lays down a layer of body fat. The brain grows billions of neurons.

Birth

28   29   32   33   34   35   36   37   38   39   40

By week 34, the baby can distinguish between dark and light. The baby is about 17 inches long and weighs about 3 pounds.

By week 39 the pregnancy is considered full term. The lungs are able to function properly outside the womb. The baby can grasp objects.

By week 40 the baby is ready to be born. The average size of a newborn is 20 inches long and 7.5 pounds.

FIGURE 7.4   Three trimesters of pregnancy and fetal milestones.

Relaxin Causes a Generalized Increase in Flexibility During Pregnancy

There is general consensus in the literature that joint laxity increases during pregnancy, and this been shown to be associated with various musculoskeletal disorders. Metacarpophalangeal (relating to the joints at the base of the fingers) and generalized laxity is thought to increase considerably at the second trimester of pregnancy (Cherni et al. 2019). Around 45 percent of all pregnant women and 25 percent of all postpartum women suffer from pregnancy-related pelvic girdle pain and pregnancy-related low back pain (Wu et al. 2004). According to Carvalho and colleagues (2017), low back pain is also more frequent in the second trimester of pregnancy. Pain in the hand and wrist is the second most prevalent musculoskeletal symptom during pregnancy (Nygaard et al. 1989).

However, it is not so clear what role the pregnancy hormones play in this. A review by Dehghan and colleagues (2014) on the effect of relaxin on the musculoskeletal system highlighted conflicting evidence for the role that relaxin plays in increased joint laxity. The review stated that the role of this hormone on the human pubic symphysis (the junction between the two hip bones at the front of the pelvis) is unknown. A systematic review by Aldabe and colleagues (2012) stated that a direct relationship between high levels of relaxin and increased pelvic mobility or peripheral joint mobility in pregnant women has not been shown. Marnach and colleagues (2003) stated that while peripheral joint laxity increases during pregnancy, these changes do not correlate well with maternal estrogen, progesterone, or relaxin levels.

Each pregnancy is unique, but there are some general rules for practicing yoga during pregnancy that most experts agree on. If a student has never practiced yoga or has practiced very little before her pregnancy, it is recommended that she practice only prenatal yoga while pregnant. If she already had a strong yoga practice before her pregnancy, she may be able to continue her typical practice with modifications particularly after the first trimester. Many yoga studios and yoga teachers encourage pregnant women who are in their first trimester to rest due to a higher risk of miscarriage during this time. However, in a systematic review and meta-analysis, Davenport and colleagues (2019) concluded that prenatal exercise is not associated with increased odds of miscarriage or the number of stillbirths and deaths in the first week of life. In plain terms, this suggests that exercise, including yoga, is safe for the unborn child. However, because so many anatomical and physiological changes occur during pregnancy, specialist prenatal and postnatal training is essential if you wish to teach yoga to pregnant women.

Menopause

Menopause is defined as the permanent cessation of ovarian function and is thereby the end of a woman's reproductive phase (Sherman 2005). Menopause begins around the

Can Yoga Have an Impact on Pregnancy, Labor, and Birth Outcomes?

In a 2012 systematic review of the literature on yoga for pregnant women, Curtis, Weinrib, and Katz (2012) included six trials: three randomized controlled trials and three controlled trials. They concluded that yoga is well indicated for pregnant women and leads to improvements on a variety of pregnancy, labor, and birth outcomes. The authors noted that regardless of the type of yoga or specific postures used, modifications should be made according to the specific needs of the individual woman in the prevention of overexertion, stress on the fetus, and premature labor. They commented that yoga is a low-impact, easily modifiable, and mindful activity, and they considered it to be a safe and sustainable activity for pregnant women. The authors also suggested that further randomized controlled trials are needed to provide more information regarding the utility of yoga interventions for pregnancy.

Riley and Drake (2013) carried out a systematic review of the literature on the effects of prenatal yoga on birth outcomes. The authors included both controlled and qualitative studies due to the limited body of research, concluding that all studies found that prenatal yoga provided significant benefits and that no adverse effects were reported. Significant findings from the randomized studies included an increase in infant birth weight, lower incidence of pregnancy complications, shorter duration of labor, and less pain among yoga practitioners. Significant findings from the nonrandomized and qualitative studies included decrease in pain, improved quality of sleep, increased maternal confidence, and improved interpersonal relationships among pregnant women who practiced yoga.

A review by Kinser and colleagues (2017) reviewed 15 studies of physical activity and yoga-based approaches for pregnancy-related low back and pelvic pain. Although additional research is required, the review suggested that nonpharmacologic treatment options, such as gentle physical activity and yoga-based interventions for pregnancy-related low back and pelvic pain and related symptoms can be recommended.

In addition to these benefits, a study of 335 pregnant women by Narendran and colleagues (2005) looking at the efficacy of yoga on pregnancy outcomes concluded that an integrated approach to yoga during pregnancy decreases intrauterine growth retardation either in isolation or associated with pregnancy-induced hypertension, with no increased complications.

Although past intervention studies have looked at yoga's effect on sleep, the first and only controlled study demonstrating the effects of mindfulness-based interventions on sleep quality in pregnant women was completed by Beddoe and colleagues (2010). They found that no significant sleep improvements were shown with a yoga intervention. However, women who began the intervention in their second trimester had fewer awakenings and less awake time during the night compared to women who began a yoga intervention in their third trimester.

It is important to note that hot yoga is not recommended at any time during pregnancy because increased core temperature is associated with an increased risk for birth defects (Duong et al. 2011).

age of 50 years and is characterized by at least 12 months of amenorrhea (the absence of menstruation) (Gracia et al. 2005). While it is an inevitable part of every woman's life, about three out of every four women experience complaints during menopause, the most common including hot flashes, night sweats, fatigue, pain, decreased libido, and mood changes. These symptoms often persist for several years after menopause (Cramer et al. 2012).

During the menopausal transition period, the drop of estrogen leads to more bone resorption than formation, which can lead to osteoporosis. Osteoporosis is a systemic skeletal condition characterized by low bone mass and microarchitectural deterioration of bone tissue that increases bone fragility and risk for fractures (U.S. Department of Health and Human Services 2004). The major health threat of osteoporosis is osteoporotic fractures. The prevalence of osteoporosis and related fractures increases in postmenopausal women (Ji and Yu 2015). Menopause is also a risk factor for cardiovascular disease because estrogen withdrawal has a detrimental effect on cardiovascular function and metabolism (Rosano et al. 2007).

In a paper discussing perspectives on menopause across different continents, Baber (2014) writes:

> There is a school of thought that believes that menopausal symptoms are a peculiarly "Western" phenomenon, not experienced by women from other regions and particularly not from Asia where, it has been claimed, dietary, social, and cultural factors afforded protection for women living in that region. More recently, studies conducted in multi-ethnic communities living in Western countries as well as in Asian communities have found that the menopause and its consequences are similar world-wide. Ethnic differences within Asia account

Can Yoga Improve Menopausal Symptoms?

There are conflicting reports from the reviews that have been conducted on the topic of whether yoga can improve the various symptoms of menopause. A systematic review by Lee and colleagues (2009) concluded that yoga is ineffective in relieving any menopausal symptoms including psychological symptoms. A systematic review and meta-analysis by Cramer and colleagues (2012) found moderate evidence for short-term effectiveness of yoga for psychological symptoms in menopausal women. However, no evidence was found for improvements regarding somatic, vasomotor, urogenital, or total menopausal symptoms. Further, no group difference was found when comparing yoga to exercise. A qualitative systematic review on mind–body interventions concluded that there was moderate evidence that yoga might relieve common menopausal symptoms including vasomotor and psychological symptoms (Innes, Selfe, and Vishnu 2010). The most recent systematic review and meta-analysis by Cramer, Peng, and Lauche (2018) reported that yoga seems to be effective and safe for reducing psychological, somatic, vasomotor, and urogenital menopausal symptoms. They also noted that the effects were comparable to those of other exercise interventions. So, it appears that over the last decade, as the body of research on this topic has slowly grown, the benefits associated with practicing yoga for menopausal symptoms have become clearer.

for small differences in endogenous hormone levels and age at menopause between Asian and Western women, and the type of menopause symptoms and their prevalence also differ between those two communities. However, like in the West and perhaps because of a Western influence, the long-term health problems of postmenopausal women including cardiovascular disease, osteoporosis and breast cancer are of major importance to Asian women and health services in the 21st century. (page 23)

MALE REPRODUCTIVE SYSTEM

The structures of the male reproductive system (figure 7.5) include the testes, the penis, and the ducts and glands that produce and carry semen.

The testes are the male reproductive organs, and they are responsible for the production of sperm and testosterone. Their location away from the pelvis is important because sperm production occurs more efficiently at a temperature lower than core body temperature. Approximately 100 to 300 million sperm are produced each day.

Sperm are transferred to the epididymis, a highly convoluted duct behind the testis, where they mature. It can take days for sperm to pass along the length of the coiled epididymis; they finally exit during an ejaculation via the ductus deferens. Sperm make up only 5 percent of the final volume of semen—the thick, milky fluid that the male ejaculates. The seminal vesicles and prostate gland add fluids to the sperm to create semen.

The prostate gland, which is unique to the male reproductive system, sits anterior to the rectum at the base of the bladder surrounding the urethra. It is typically the size of a walnut in adulthood and is formed of both muscular and glandular tissues. It excretes an alkaline, milky fluid that is critical to first coagulate and then decoagulate the semen following ejaculation. The prostate normally doubles in size during puberty, and at approximately age 25, it gradually begins to enlarge again.

The final addition to semen is made by two bulbourethral glands that release a thick, salty fluid that lubricates the end of the urethra and the vagina and helps to clean urine residues from the penile urethra.

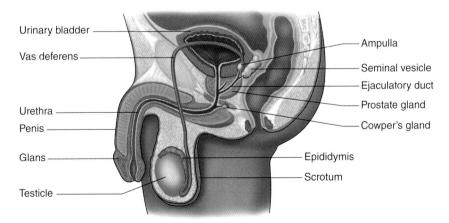

FIGURE 7.5 Male reproductive system.

The penis is the male organ of copulation; it is composed of columns of erectile tissue called the *corpora cavernosa* and *corpus spongiosum*, which fill with blood when sexual arousal activates vasodilatation in the blood vessels of the penis.

CONDITIONS OF THE REPRODUCTIVE SYSTEM

We will now discuss some of the most common conditions that affect the female and male reproductive systems and, where possible, explore the effect that yoga may have on these conditions.

Premenstrual Syndrome

Premenstrual syndrome (PMS) is characterized by physical, psychological, and behavioral changes and is believed to affect 75 percent of women of childbearing age (Zaafrane et al. 2007). Matsumoto and colleagues (2007) suggested that functioning of the autonomic nervous system during the late luteal phase of the cycle is altered, which could be associated with diverse psychosomatic and behavioral symptoms appearing during the premenstrual phase.

Dysmenorrhea

Primary dysmenorrhea, or menstrual cramps, is one of the most prevalent gynecologic conditions, affecting an estimated 67 to 90 percent of young women (Ju, Jones, and Mishra 2014). Occurring prior to or during menses, it is described as sharp, cramping,

Can Yoga Alleviate PMS Symptoms?

A study by Kanojia and colleagues (2013) looked at the effect of yoga on autonomic functions and psychological status during the premenstrual and postmenstrual phases of the menstrual cycle in young healthy females. Fifty participants between the ages of 18 and 20 were randomized into two groups. One group consisted of subjects who practiced yoga 35 to 40 minutes per day, six times per week for the duration of three menstrual cycles. The second group acted as control subjects.

The authors reported that regular practice of yoga beneficially affected the premenstrual phase of the cycle by improving parasympathetic activity and bringing equanimity of mind. Further studies are required with longer time periods and larger sample sizes to expand upon this interesting insight. The results of a randomized controlled trial by Kamalifard and colleagues (2017) also show that yoga significantly reduces the symptoms of PMS and can be prescribed for treatment of PMS. Ghaffarilaleh, Ghaffarilaleh, Sanamno, and Kamalifard (2019) found that yoga positively affected depression and blood pressure in women with PMS in a randomized controlled clinical trial. Ghaffarilaleh, Ghaffarilaleh, Sanamno, Kamalifard, and Alibaf (2019) looked at the effects of yoga on the quality of sleep of women with PMS. They concluded that yoga reduced the disturbances of sleep for women with PMS, which subsequently improved the efficiency of their sleep. Therefore, yoga can be prescribed for improving sleep disturbances in women with PMS.

Can Yoga Alleviate Symptoms of Dysmenorrhea?

A systematic review by Ko, Le, and Kim (2016) of randomized controlled trials looked at the effects of yoga on dysmenorrhea. At the time, there was only evidence from two randomized controlled trials that yoga interventions may be effective for dysmenorrhea. Therefore, further high-quality randomized controlled trials are required to investigate the hypothesis that yoga alleviates menstrual pain and the symptoms associated with dysmenorrhea, to confirm and further comprehend the effects of standardized yoga programs in dysmenorrhea. In a systematic review, McGovern and Cheung (2018) also reported that yoga can mitigate menstrual pain from primary dysmenorrhea, reduce the social and psychological distress associated with the condition, and function as a potential quality-of-life improvement method without adverse side effects. The authors included 14 studies in their review and suggested that future research using larger randomized controlled trials of high methodological quality was needed to ascertain the magnitude of yoga's clinical significance. Kim (2019) completed a meta-analysis of randomized controlled trials, which looked at the effect of yoga on menstrual pain in primary dysmenorrhea. They included four trials and concluded that yoga is an effective intervention for alleviating menstrual pain in women with primary dysmenorrhea.

The underlying mechanism of yoga's quality-of-life improvement potential is not fully understood. However, evidence indicates it may relate to physiological, behavioral, and psychological changes consequent to participation in a yoga practice, including endorphin release, increased parasympathetic nervous system activity, and downregulation of the sympathetic nervous system and hypothalamic-pituitary-adrenal axis (McGovern and Cheung 2018; see chapters 2 and 6).

---

or gripping pain felt in the lower abdomen, backs, or thighs, and lasts from hours to days. It is not associated with an underlying pathology (Dawood 2006), but women with primary dysmenorrhea report significantly lower quality of life because of pain, general health condition, and physical and social functioning (Iacovides, Avidon, and Baker 2015).

Infertility

Female infertility is generally defined as not being able to become pregnant after one year or more of unprotected sex. The National Center for Health Statistics (2019) reported that approximately 9 percent of women aged 15 to 49 in the United States are infertile. About one-third of infertility cases are caused by fertility problems in women, and another one-third of fertility problems are due to fertility problems in men. The other cases are caused by a mixture of male and female problems or by problems that cannot be determined.

Two of the most common fertility treatments are intrauterine insemination and in vitro fertilization. In intrauterine insemination, healthy sperm is collected and inserted directly into the uterus at the time of ovulation. In in vitro fertilization, oocytes are taken from the ovaries and fertilized by sperm in a laboratory, where they develop into embryos. The embryo is then transferred to the uterus.

The experience of fertility treatment can be daunting and stressful. Yoga can potentially help couples overcome infertility and increase the success rate of fertility treatments by improving the physiological and psychological states of both men and women. In a review by Miner and colleagues (2018) looking at the evidence for the use of complementary and alternative medicines during fertility treatment, the authors included three studies that evaluated the use of yoga and fertility outcomes. They stated that all three studies showed an improvement in mental health outcomes, which may translate to a decreased drop-out rate during the fertility treatment. In a prospective study by Oron and colleagues (2015), the authors concluded that anxiety, depression, and fertility-specific quality of life showed improvement over time in association with participation in a six-week yoga program in women awaiting their treatment with in vitro fertilization. Gaitzsch and colleagues (2020) looked at the effect of mind–body interventions, including mindfulness-based interventions and yoga, on psychological and pregnancy outcomes in infertile women. The authors concluded that their review of 12 studies offers evidence for the effectiveness of mind–body interventions in

TRY IT YOURSELF: Mindfulness Meditation

Mindfulness is simply being fully present in a given moment, aware of where we are and what we are doing. Mindfulness can be a formal meditation practice, or it can be incorporated into the simple tasks that we perform throughout each day. When we are mindful of our actions, we become more tuned into our senses, thoughts, and emotions.

Start by finding a comfortable place to practice that offers little in the way of distractions, and then find any comfortable position—seated or lying down. Allow yourself to settle into the space. Close your eyes if that feels accessible to you today; otherwise soften your gaze and focus on a fixed object down in front of you. Release your jaw, let your lips and teeth gently part, and allow your tongue to fall away from the roof of your mouth. Begin to take a couple of gentle breaths in and out through your nose, noticing the parts of your body that naturally move as you inhale and exhale. Draw your attention to the air moving through your nostrils. With each breath, you can mentally note breathing in and breathing out. When you inevitably notice your mind wandering, simply return your attention to your breath. Instead of judging yourself for this, practice observing without the need to react. Before making any physical adjustments to your position, pause for a moment and then, with intention, make any movements you need to. This practice is basically just about being still and paying attention to what arises, without judgment or expectation. As simple as that might sound, this can be really challenging and takes a lot of time and patience. When you are ready to draw the practice to a close, either slowly open your eyes or gently lift your gaze. Notice any sounds in the environment around you and how your body feels right now. Notice your thoughts, feelings, and emotions. You can now choose to take a moment to set an intention for the remainder of your day.

reducing anxiety state and depression in infertile women and a possible improvement in pregnancy rate. In addition to what the aforementioned studies found, it is also probably safe to assume that yoga, when practiced as a couple, could also provide some important bonding time between the two partners during fertility treatment, which can be very emotionally draining.

Prenatal and Postnatal Depression

Depression has been estimated to affect approximately 10 to 15 percent of pregnant women, while, specific to the postpartum period, close to 20 percent of women meet criteria for major or minor depression within the first three months of childbirth (Gavin et al. 2005). Prenatal depression is associated with an increased risk of the offspring having emotional, behavioral, and cognitive difficulties. Postnatal depression adds further risks to mothers' health, parenting, and child development (Stein et al. 2014).

In a systematic review and meta-analysis exploring the effectiveness of yoga for prenatal depression, Gong and colleagues (2015) concluded that prenatal yoga intervention in pregnant women may be effective in partly reducing depressive symptoms. In a randomized controlled trial by Buttner and colleagues (2015), 78 percent of women practicing yoga postpartum experienced clinically significant changes in depressive symptoms.

Male Reproductive Health and Yoga

Sperm DNA damage is a common underlying cause of male infertility, recurrent spontaneous abortion, recurrent implantation failure, and congenital malformation (Gautam et al. 2018). Around 60 percent of infertile men, especially those with normal sperm, have high seminal free radical levels and low antioxidant levels. This is now believed to be the major cause of defective sperm function. Yoga and meditation practices can result in improvement in standard sperm parameters, but they are also ideal in treating oxidative stress and oxidative DNA damage (Dhawan et al. 2019). This strategy may not only reverse testicular aging but also result in overall improvement in the health and quality of life of such men and those of the next generation. Yoga may also help to promote a significant decline in oxidative DNA damage and normalization of sperm transcript levels. This may not only improve pregnancy outcomes but also improve the health trajectory of the offspring (Dhawan et al. 2018).

Irritable Male Syndrome

While it might sound like a joke, *irritable male syndrome* is a term coined by psychotherapist and author Jed Diamond (2004) to describe male hormonal fluctuations and the symptoms they may cause. Diamond based his theory on animal research by Lincoln (2002), who studied rams. Diamond suggested that the symptoms of irritable male syndrome mimic some of the symptoms that women experience during premenstrual syndrome. However, there is no medical evidence of this syndrome, and it is not a recognized medical diagnosis.

CONCLUSION

It is reassuring to know that yoga can have a positive impact on the female and male reproductive systems, leading to improvements on a variety of pregnancy, labor, and birth outcomes and potentially improving symptoms of premenstrual syndrome and menopause. Yoga can also play an important role in managing stress, anxiety, and depression during fertility treatment.

Chapter 8

DIGESTIVE SYSTEM

Through digestion, we absorb the outside world. Through our intestines, that which is outside us becomes us. Our bodies need nutrients from food and drink to work properly and stay healthy. Through the digestive process, our bodies absorb proteins, fats, carbohydrates, vitamins, minerals, and water. Our digestive systems break nutrients into parts that are small enough to be transported by our blood to be used by every cell in our bodies for energy, growth, and cell repair. Proteins are broken into amino acids. Fats are broken into fatty acids and glycerol. Carbohydrates are broken into simple sugars. The foods we eat can have a major impact on how we feel, perform, and simply exist. While choosing to engage in exercise like yoga is optional, ingesting food is requisite for life.

Yoga has a special relationship with diet and the digestive system. While many yoga lineages prescribe specific diets, it is often claimed that yoga can ignite the digestive fire or squeeze and soak the internal organs like a sponge while twists help the liver detoxify. But is there any truth to these claims? First, we need to understand a little bit about the structures and functions of the digestive system.

ANATOMY AND PHYSIOLOGY OF THE DIGESTIVE SYSTEM

The digestive system (figure 8.1) is made up of the gastrointestinal tract (GI tract)—also called the *digestive tract*—and the liver, pancreas, and gallbladder. The GI tract, which can be up to 30 feet (over 9 m) in length in adults, consists of a series of hollow organs joined in a long, twisting tube from the mouth to the anus. The hollow organs of the GI tract are the mouth, the esophagus, the stomach, the small intestine, the large intestine, and the anus; the liver, pancreas, and gallbladder add secretions to aid digestion. All these organs serve to break down the food and drink we consume to obtain energy and nourishment. These organs together perform six tasks: ingestion, secretion, propulsion, digestion, absorption, and defecation.

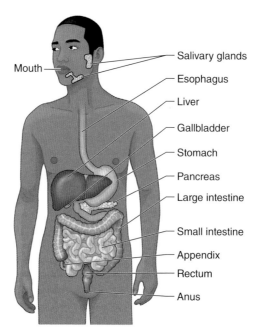

FIGURE 8.1 Digestive system.

Mouth and Esophagus

The process of digestion starts with the mouth where food is mechanically broken down and mixed with saliva, which contains enzymes that aid digestion. One such enzyme is salivary amylase, which breaks starch, a complex carbohydrate, into simple sugars. This is why starchy foods like popcorn or potatoes start to taste sweet after about 30 seconds in the mouth as salivary amylase turns the starch into sugar. The other enzyme released in the mouth is lingual lipase, which begins the process of breaking down fat into simpler fatty acids. Hence, the mouth watering we experience upon reading a menu or smelling food is a physiological response of increased saliva production to prepare the body for digestion.

Taste, or gustation, is a form of chemoreception, or sensing of chemicals. Specialized taste receptors contained within our taste buds are stimulated by the chemicals of food and send messages to the brain, which then differentiates the tastes as pleasant or harmful and creates the perception of taste. Taste buds are mainly situated on the top of the tongue but also present on the epiglottis and upper part of the esophagus—the food pipe. The brain can differentiate between the many chemical qualities of food with the five basic tastes referred to as *saltiness*, sourness, bitterness, sweetness, and umami. Sensing saltiness helps ensure our bodies have the right level of sodium, an essential nutrient, while sweetness guides us to the foods that will supply energy, hence the very common love of sweet foods is biologically driven and once was of great benefit to ensure we could get enough calories to survive. Sourness is a result of a food's acidity, while bitterness is often caused by toxic compounds, thus creating a generally unpleasant taste to which many are averse. The perception of bitterness helps prevent us from ingesting poisonous substances, because many toxins, including those found in food that has gone bad, taste bitter. Many bitter foods are very good for us, including dark leafy greens, parts of citrus fruits, berries, coffee, tea, and cocoa. Finally, umami is thought to signal food high in protein.

Olfaction, or sense of smell, also plays an important role in our perception of taste. Olfactory receptors, which were discovered and first described in 1991, are chemoreceptors like taste receptors. Located on cell surfaces in the nose, olfactory receptors bind to chemicals enabling our detection of smells. Decreased olfaction manifests as taste loss (Pinto 2011). Anyone who has eaten while having a blocked nose, perhaps as a result of a common cold, has probably experienced this. Beyond the inevitable cold, Vennemann, Hummel, and Berger (2008) found that smoking significantly increases the risk of impairment of olfactory function—yet another adverse health effect of smoking. From a large-scale survey of U.S. adults, Hoffman, Ishii, and MacTurk (1998) found that the prevalence of chemosensory impairment—in other words, a reduced ability to smell and taste—increased with age. This had farther-reaching consequences than simply finding less flavor in food. The researchers found that chemosensory disorders were associated with functional limitations (including difficulty standing or bending), depression, phobias, and several other health-related characteristics. Indeed, taste is another part of the puzzle that makes us the interconnected, complex organisms we are.

After the food has been tasted and the brain has decided to ingest it, the tongue then pushes food into the throat. A small, leaf-shaped flap of elastic cartilage, the epiglottis, folds over the windpipe to prevent choking, and the food enters the esophagus, a muscular tube connecting the mouth to the stomach (figure 8.2). Muscles in the esophagus contract and relax in a wavelike manner, called *peristalsis*, pushing

How to Practice Yoga While Eating

Mindful eating could be described as an intersection between the mindfulness of yoga and the act of eating. Mindful eating is a practice of devoting full attention to one's experiences, cravings, and physical cues when eating. Mindful eating involves eating slowly and without distraction, listening to physical hunger cues and eating only until you are full, distinguishing between genuine hunger and nonhunger triggers for eating as well as engaging your senses by noticing colors, smells, sounds, textures, and flavors. The practice can also go beyond the kitchen table to notice the effects food makes you feel afterward. While many of these practices sound like common sense, Americans spend less time eating than in previous decades (Zeballos and Restrepo 2018), families spend almost half of mealtime distracted by technology or other things (Saltzman et al. 2019), and obesity is growing to epidemic prevalence. Mindfulness, however, allows you to replace automatic thoughts and reactions with more conscious, healthier responses (Sears and Kraus 2009). In fact, people practicing mindful eating have been observed to consume fewer calories per meal and to choose healthier foods (Jordan et al. 2014). Practicing mindful eating might be a great way of bringing one's yoga practice to the dinner table.

food along the esophagus and into the stomach. A ringlike muscle at the end of the esophagus, called the *esophageal sphincter*, controls the passage of food into the stomach. Because the movement of the esophagus and this sphincter is well coordinated, it is possible to eat while upside down, though it is probably not recommended. If the esophageal sphincter relaxes at the wrong time, stomach acid can move up to the esophagus, causing acid reflux, or heartburn. The diaphragm plays a role in preventing acid reflux by applying pressure to the esophagus. Slouching while eating can contribute to this occurrence.

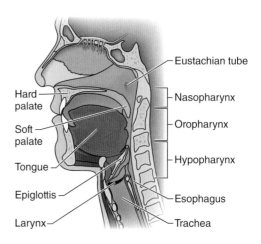

FIGURE 8.2 Mouth, tongue, epiglottis, trachea, and esophagus.

Similarly, acid reflux is very common in the late stages of pregnancy as the uterus presses up on the stomach.

Stomach

The stomach is distensible, meaning that it can expand to its contents, able to hold about one liter of food owing to the stomach lining's many folds. The stomach of a newborn baby, however, can expand to hold only about 30 milliliters, 33 times less than an adult, and hence the need for frequent feeds. The stomach liquefies food and adds digestive acids with hydrochloric acid being the main constituent. The acids produced by our stomach lining are so strong (pH of 1-3) that they would cause a chemical burn to our skin on contact. But the lining of the stomach produces a mucus layer that protects it, preventing the stomach from digesting itself. Other cells called *parietal cells* in the stomach lining produce bicarbonate, a base, and release it as needed to attain just the right level of pH. (Over-the-counter antacids for the relief of heartburn also contain carbonates that affect stomach pH.) The gastric acids break down the ingested fats and the long chains of proteins into their simpler amino acids. In addition, many microorganisms are inhibited or destroyed in an acidic environment, preventing infection or sickness.

At the bottom of the stomach is the pylorus, containing many glands that secrete digestive enzymes. After spending one to two hours in the stomach, the food has become a thick semiliquid called *chyme*, which then exits the pylorus via the pyloric sphincter and is slowly released into the small intestine.

Intestines

From the pyloric sphincter, chyme enters the first part of the small intestine, the duodenum, where it mixes further with digestive enzymes from the pancreas, and then continues its journey through the middle of the small intestine, the jejunum, and finally, the ileum. Millions of tiny fingerlike projections called *villi* line the walls of the small intestine and absorb nutrients as the contents are moved along with peristalsis. These villi act to increase the surface area of the small intestine area by 60 to 120 times, thus increasing the small intestine's ability to absorb nutrients. Helander and Fändriks (2014) calculated the total surface area of the average small intestine to be around the size of half a badminton court! Secretions from the liver, pancreas, and gallbladder are also emptied into the small intestine to aid in the digestive process. Once the chyme is fully broken down to its constituent parts, it is absorbed into the blood through the villi. Around 95 percent of nutrient absorption occurs in the small intestine.

After traveling through nearly 20 feet (6 m) of small intestine, unabsorbed material moves into the large intestine where the remaining liquids and salts (also called *electrolytes*) are absorbed. The large intestine includes the cecum, the colon, and the rectum, as well as the appendix, a finger-shaped pouch attached to the cecum. Within the large intestine, bacteria further break down the undigested material, which continues to solidify and eventually passes from the body as feces through the rectum and anus during a bowel movement. In the large intestine, some vitamins, such as biotin and vitamin K produced by bacteria in the gut flora of the colon, are also absorbed.

Our Internal Ecosystem

There is another important element to our digestive system, beyond the organs named so far. If we count all the cells in and on the body, both human and foreign, our bodies are more foreign than human. On average, our bodies are composed of about 43 percent human cells and 57 percent microorganisms (Sender, Fuchs, and Milo 2016). Those numbers are of course averages, so one person might have half as many microbiota, while another might have twice as many. It is certain that we are not just the cells created from our DNA code but also a host for many microscopic colonists, some of which are essential to good health.

The human microbiome is the aggregate of all the microorganisms that reside on or within all our tissues, organs, and fluids. These microbiota dwell and thrive on your skin, lungs, saliva, eyes, biliary tract (including the liver), seminal fluid, placenta, uterus, and ovarian follicles, among other organs. By far, though, the GI tract has the largest numbers of bacteria and the greatest number of species with literally trillions of bacteria at any given time.

Our whole-body microbiome begins forming when we are just a fetus in the womb when we are exposed to our mother's bowel and uterine microbiota (Younge et al. 2019). A baby traveling down the vaginal canal during birth is exposed to many beneficial microorganisms from the mother, rapidly colonizing and strengthening the newborn's immunity and GI tract. Shao and colleagues (2019) found that babies delivered via cesarean section had disrupted levels of beneficial bacteria from their mother and higher levels of opportunistic pathogens associated with the hospital. They found these effects were also seen, to a lesser extent, in vaginally delivered babies whose mothers had been taking antibiotics and in babies who were not breastfed immediately following birth. This study, the largest of its kind, also revealed, perhaps counterintuitively, that the microbiome of vaginally delivered babies did not come from the mother's vaginal bacteria but from her gut bacteria. It is not known whether this difference in microflora affects people later in their lives, but the researchers did find that differences in gut bacteria between vaginally born and cesarean-delivered babies largely evened out by one year old. Larger follow-up studies are needed to determine if the early differences influence later health outcomes. What this study does show is the complexity of the human microbiome and how little we understand about it. The authors point out that cesarean sections are often lifesaving procedures, so women should not be put off by the idea of a cesarean section because of this study.

Adults generally have greater diversity of microbiota than children, although interpersonal differences are greater in children than in adults (Yatsunenko et al. 2012). The microbiota mature into an adultlike configuration mostly during the first three years of life (Yatsunenko).

Many bacteria have a commensal relationship with their human host, meaning that they neither harm nor, as far as we know, benefit the host but simply exist and feed off the materials in the gut. Some bacteria, though, offer a mutualistic relationship wherein both benefit. For example, certain bacteria ferment dietary fiber into short-chain fatty acids, which can then be absorbed by the host. Intestinal bacteria also play an important role in synthesizing vitamin B and vitamin K as well as metabolizing byproducts of digestion like bile acids, sterols (like cholesterol), and xenobiotics (like pollutants and drugs).

In this way, the gut flora appears to function like another organ of the body, and irregularities of the flora, which is called *dysbiosis*, have been correlated with a variety of inflammatory and autoimmune conditions. Inflammatory bowel disease, which consists of ulcerative colitis and Crohn's disease, is connected to dysbiosis, or disruptions in the gut microbiota. This dysbiosis presents in the form of decreased microbial diversity in the gut and is correlated to defects in host genes, thus changing the natural immune response in individuals (Hold et al. 2014). Dysbiosis has even been linked to neural diseases such as autism and Parkinson's disease.

Researchers have found that long-term psychological stress is linked with dysbiosis (Qin et al. 2014). Some researchers have also hypothesized that changes to our gut microbiota are connected to our rigorous hygiene, increased consumption of processed foods, and wide use of antibiotics. While rigorous hygiene and antibiotics have their uses and can even be lifesaving, all these factors might contribute to reduced genetic diversity of human microbiomes in the developed world.

In a remote village of the Amazon, a small, isolated tribe, the Yanomami people, were found to have the most diverse microbiota of any humans yet. Researchers analyzed the microbial DNA in oral, fecal, and skin samples of 34 Yanomami, finding their average microbiota had twice as many genes as that of the average U.S. person (Clemente et al. 2015). The Yanomami's microbiomes are even more diverse than those of indigenous tribes in Africa. However, more research is needed to properly understand gut microbiota and what effect a less diverse microbiome might have.

We all have a unique blend of gut flora, which is influenced by our genes, our diet, our age, our stress levels, and environmental factors, among many other variables, creating something like a gut fingerprint (figure 8.3). This might be one of the reasons two people can eat the same thing and have different digestive experiences.

Another factor that seems to affect the gut microbiota is exercise. Matsumoto and colleagues (2008) found that rats that exercised in the form of voluntary wheel running

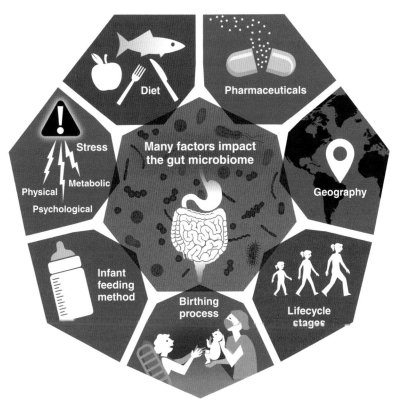

FIGURE 8.3 Factors that affect gut flora include genetics, age, birth delivery route, antibiotic or drug use, diet, and exercise.

had increased growth of bacteria that produced the fatty acid, butyrate. Butyrate can promote repair of the gut lining and reduce inflammation, potentially preventing diseases such as inflammatory bowel disease and insulin resistance, which can lead to diabetes. While effects on humans might be different than those on rats, both are mammals sharing much of the same DNA, so rodent studies can offer a suggestion of what is also happening in humans. Exercise can also create microbiome changes that guard against obesity and improve metabolic function. Even a modest physical activity regimen can affect the microbiome. Bressa and colleagues (2017) found that women who performed at least three hours per week of light exercise, such as a brisk walk or swim, had increased levels of three specific types of bacteria that promote health by reducing inflammation or improving metabolism; sedentary subjects were associated with less microbiota richness. Additionally, athletes have very different microbiota profiles compared to sedentary people of similar age and sex. Athletes generally have more diverse microflora and a higher abundance of the three beneficial bacteria mentioned earlier (Mohr et al. 2020). While there has not been any published research about the effect of yoga on the microbiota, it seems that exercise, of which yoga is one form, does affect our gut microbiome in a positive way.

What is clear is that the health of our gut microbiome is an important factor in the health of our whole digestive system and our whole organism. Looking after ourselves also has implications in looking after our microscopic colonists. While there is much more to learn about the gut flora, we should not underestimate how these microscopic organisms shape us.

Liver

Second only to the skin, the liver is one of the largest organs in the body and is an accessory digestive gland that plays a role in metabolism. The liver is located in the upper right quadrant of the abdomen just below the diaphragm, to which it is attached at one part. It is to the right of the stomach, and it overlies the gallbladder.

The liver has many different functions. It synthesizes proteins, produces biochemicals needed for digestion, and regulates the storage of glycogen, which it forms from the glucose found in food or already in the bloodstream, thus assisting in regulating blood glucose levels.

Our body has a constant need for energy, but we are not constantly eating. The liver absorbs glucose from food while we eat and stores it (as glycogen). Then it mobilizes stored glucose to be used for energy in the cells. If we then have an excess of glucose in the blood, the liver will again absorb the glucose for later use. In this way, the liver plays a very important role in regulating a constant level of blood glucose.

But the liver is most popularly known for its detoxification function, breaking down various metabolites—both those produced within the body, such as lactic acid, and those consumed, such as alcohol.

 **Twists Help the Liver Detoxify**

B.K.S. Iyengar, student of Tirumalai Krishnamacharya and creator of Iyengar Yoga, famously stated in his classes that deep twists cause a squeeze-and-soak effect on the body, especially the intervertebral discs and the internal organs. Whether Iyengar was the original source of this idea or whether it started from someone before him, the idea of twists detoxing the body has certainly become commonplace and is firmly planted in the minds of many yoga practitioners. An article published in 2008 in *Yoga Journal* attempts to explain Iyengar's idea:

> *The theory is that twists cleanse the internal organs in much the same way that a sponge discharges dirty water when squeezed and can then absorb fresh water and expand again. The idea is that, when you twist, you create a similar wringing action, removing stale blood and allowing a freshly oxygenated supply to flow in. (Rizopoulos 2017, para 2)*

While Iyengar, who died of kidney failure in 2014, may have been well meaning in his assertions about the physiology of detoxification, the squeeze-and-soak theory is inaccurate and has no scientific basis. Every second of every day, whether you are asleep or awake, your body is constantly detoxifying itself, and this detoxification is a cellular process, not a mechanical one that requires the compressing, wringing, or stretching of internal organs.

Your body is regularly exposed to toxins, poisonous substances produced by living organisms, including those produced in the body itself, such as lactic acid and microbial waste products in the gut. Your body removes these toxins through the liver, feces, and urine. The liver, specifically, alters toxic substances chemically, yielding them harmless and ready for excretion.

Even with our bodies' integrated detoxification system, there are still some chemicals that cannot be easily removed through these processes, including persistent organic pollutants (found in pesticides), phthalates (found in hundreds of plastic products), bisphenol A (found in many food containers and hygiene products), and heavy metals (found in agriculture, medicine, and industry). Known to accumulate in the body and taking a very long time, potentially years, to be removed, these chemicals may be linked to various chronic diseases including asthma, cancer, autism, and attention-deficit/hyperactivity disorder (Sears and Genuis 2012).

The idea that a certain yoga pose, a certain diet, or a certain product can remove these stubborn toxins from the body is a tempting one. However, there is little to no evidence that one movement, diet, or product can do just that. In a 2015 review of all studies to date examining the efficacy of detox diets, Klein and Kiat found that we have very little clinical evidence to support the use of detox diets despite a booming detox industry and product packaging that makes bold but unsubstantiated claims. Although a small number of studies have found commercial detox diets might enhance liver detoxification and eliminate persistent organic pollutants, these studies had flawed methodologies and small sample sizes, thus reducing their scientific credibility. Klein and Kiat concluded:

> To the best of our knowledge, no randomised controlled trials have been conducted to assess the effectiveness of commercial detox diets in humans. This is an area that deserves attention so that consumers can be informed of the potential benefits and risks of detox programs. (2015, p. 1).

As for yoga, there has been no research to show that it improves the body's natural detoxification system. This might be because our body already does a fantastic job of detoxification on its own. If you were not able to remove toxins from your body, you would know, as a plethora of symptoms would occur, even fatal ones.

This is not to say that yoga (or exercise in general) has no effect on our body's ability to detoxify, but the benefits likely come from exercise's ability to decrease inflammation and increase vascularity. While some inflammation is necessary in recovering from a wound or infection, chronic inflammation is hard on the body and weakens many systems. Long-term inflammation underlies many serious diseases, including cardiovascular disease and diabetes, while cancer, obesity, osteoporosis, Alzheimer's disease, and cardiovascular disease have all been linked to increased biomarkers of inflammation (Tabas and Glass 2013).

It is well established that exercise creates an anti-inflammatory response in the body—one of its most potent benefits (Flynn, McFarlin, and Markofski 2007). When inflammation is reduced, all systems of the body can work more efficiently, including those of digestion and detoxification. Another well-known benefit of exercise is increased circulation, so more blood, which contains essential nutrients and oxygen, is available for digestion and detoxification. In these ways among other benefits, exercise can certainly help optimize our body's detoxing processes.

The few studies that have examined yoga and inflammation have suggested that yoga also has the same anti-inflammatory benefits (Pullen et al. 2008; Pullen et al. 2010). Even though the research on yoga is much sparser, it is reasonable to assume that a physical yoga practice would provide many of the same exercise benefits such as decreased inflammation and increased circulation. If you are doing a lunge, your body doesn't know if you are in a yoga class or a personal training session; the physiological effects would be largely the same. Interestingly, one small study found that even a nonphysical practice of mindfulness decreased inflammation (Ng et al. 2020).

So, does yoga detoxify your body? Not in the way Iyengar suggested. Twists do not detoxify your body. Neither yoga nor diet do anything that our bodies cannot do on their own. Nonetheless, yoga can optimize the body's natural detoxification system, not through one specific pose such as a twist, but through exercise's already powerful benefits of decreasing inflammation and increasing circulation. Trust that your body is fully equipped to handle toxins and other unwanted substances.

Another thing we can do to improve our detoxification process is to simply reduce our exposure to toxins. While detox diets do not work in the way they are often advertised to, nearly all diets encourage healthful, balanced eating, which reduces your exposure to harmful products, including artificial ingredients found in junk food. We can also reduce our exposure to persistent organic pollutants by eating organically grown food. We can reduce the burden on our liver by drinking only moderate amounts of alcohol. Finally, keeping active with moderate exercise, yoga or otherwise, will help the body do what it does very well.

Gallbladder and Bile

The gallbladder is a hollow part of the biliary tract that sits just beneath the liver. The biliary tract is formed by the liver, gallbladder, and bile ducts, which all work together to make, store, and secrete bile, or gall. The gallbladder is a small organ that stores the bile produced by the liver until the bile is released into the small intestine.

Bile is produced by the liver and is an important chemical in the process of digestion. It is released from the gallbladder when food is discharged into the duodenum and for a few hours after. Bile helps in the digestion of fats by breaking down larger molecules into smaller ones in a process called *emulsification*. Bile also acts as a surfactant, which reduces the surface tension between two compounds. Once fats are dispersed by bile, they have more surface area, allowing the pancreatic enzyme lipase to break them down into smaller fatty acids and monoglycerides, which can then be absorbed by the villi lining the entire length of the small intestine. The small intestine also absorbs bile, which is then transported to the liver for reuse. Bile also plays a role in absorbing vitamin K in the diet; vitamin K is a nutrient needed for proper blood clotting and wound healing.

Gallstones are a common gallbladder pathology. Gallstones can form if there is too much cholesterol or bilirubin in the bile, or if the gallbladder doesn't empty properly. Gallstones can be very painful, though most people who have them never have any symptoms and never even know they have the stones. The recommended advice for the prevention of gallstones is, naturally, a healthy, balanced diet with plenty of fresh fruit and vegetables as well as whole grains. Some evidence shows that regularly eating nuts, such as peanuts or cashews, can help reduce the risk of developing gallstones and small amounts of alcohol may also help reduce your risk. As with most things, the key is moderation.

Pancreas

The pancreas, which lies below and at the back of the stomach, is an important organ with an accessory role in digestion, a major role as an endocrine organ, and crossover between both roles. The pancreas secretes insulin when blood sugar becomes high. The insulin moves glucose from the blood into the muscles and other tissues for use as energy. When blood sugar is low, the pancreas releases glucagon, which allows stored sugar to be broken down into glucose by the liver in order to rebalance the sugar levels. The pancreas produces large amounts of bicarbonate and secretes it to the duodenum to neutralize gastric acid coming into the digestive tract. The pancreas secretes bicarbonate ions, which are basic, to help neutralize the acidic chyme that comes from the stomach. Finally, the pancreas is the main source of enzymes for the digestion of fats and proteins. By contrast, the enzymes that digest polysaccharides, or carbohydrates, are primarily produced by the walls of the intestines.

Enteric Nervous System

When we think of the nervous system, we usually think of the brain and all the nerves that deliver information around the body. The enteric nervous system (ENS), which governs the gastrointestinal tract, is often referred to as the *second brain*. The ENS is one of the main divisions of the autonomic nervous system, which also controls other involuntary functions such as heart rate, blood pressure, and breathing rate. Through its vast web of hundreds of millions of neurons, the enteric nervous system is constantly monitoring the health of the digestive tract from the esophagus to the anus, communicating with the central nervous system (CNS) to maintain balance throughout.

Because the ENS is so complex and extensive, it can operate independently of the CNS, although they are in regular communication and influence each other. However, the ENS has been observed to be operable with a severed vagus nerve (Li and Owyang 2003), which is its main communication channel to the CNS. The ENS controls the motor functions of the digestive system and the secretion of gastrointestinal enzymes. The neurons of the ENS communicate through many neurotransmitters similar to the CNS, including acetylcholine, dopamine, and serotonin. In fact, it is estimated that more than 90 percent of the body's serotonin, a neurotransmitter believed to be associated with feelings of wellness, is produced and resides within the gut (Camilleri 2009), and altered levels of it have been linked to diseases such as irritable bowel syndrome, cardiovascular disease, and osteoporosis. Additionally, about 50 percent of the body's dopamine lies in the gut. The large presence of serotonin and dopamine in the GI tract are key areas of current research for neurogastroenterologists.

In addition to regulating the movement of food through the GI tract and the production of digestive juices, the ENS also plays an important role in communicating with the brain. The complex web of neurons, hormones, and chemical neurotransmitters within the gut sends messages to the brain about the state of the GI tract, and it receives information from the brain to directly impact the gut environment. The central nervous system can control the rate at which food is being moved and how much mucus is lining the gut, which can have a direct impact on the environmental conditions the microbiota experiences. In this way, the CNS and ENS are in constant communication about how hungry we are, whether or not we're experiencing stress, or if we've ingested a disease-causing microbe. The gut microbiota influence the body's level of the powerful neurotransmitter serotonin, which regulates feelings of wellness. Some of the most prescribed drugs for treating anxiety and depression work by modulating levels of serotonin. Serotonin is probably just one of many biochemical messengers affected by microbiota and affecting our mood and behavior. So, our emotional state and stress levels can directly impact our digestion. This gut–brain connection, which might be connected to the idea of a gut feeling, is also discussed in chapter 2 on the nervous system.

DIET AND THE DIGESTIVE SYSTEM

The digestive system is directly impacted by our diet like no other system in the body. Some diets commonly associated with yoga are intermittent fasting and the yogic diet, to include the Ayurvedic diet. What are these diets, and what effects do they have?

Intermittent Fasting

While it has become a recent diet trend, fasting has been used therapeutically since at least the fifth century BCE when Greek physician Hippocrates, considered to be the father of medicine, recommended it for patients with certain symptoms. Fasting is common to many cultures around the globe and has been practiced for millennia, often connected with religious or spiritual beliefs. In Hinduism and Jainism, followers observe a partial fast known as *Ēkādaśī* every 2 weeks on the 11th day of the lunar cycle. Many yoga traditions promote fasting, and it is considered an important part of Ayurvedic medicine. So why has this diet trend, while not new, become popular of late, and does it offer any real benefits?

The term *intermittent fasting* (IF) refers to alternating periods of fasting (or calorie restriction) and nonfasting over a given period. While fasting actually means complete abstention from food, many popular IF practices allow some food and so are more accurately called *calorie restriction diets*.

Here are three common IF methods. With the five-day-per-month approach, you eat between 700 and 1,000 calories for five days in a row each month. With the popular 5:2 plan, you eat reduced calories two days per week. On fasting days, you have 500 calories for the entire day. On nonfasting days, you still follow a healthy diet but with no calorie restriction. A third form of fasting involves not eating anything between dinner and 2 p.m. the next day, which is an accessible form as it simply lengthens the natural overnight fast, thus moving the breaking of the fast, which is the origin of the word *breakfast*, to later.

Observed Benefits of Fasting

IF is said to mimic the hunter-gatherer diet that was followed when people did not have a predictable number of calories every day, with some days being more bountiful while others were scarcer. The idea is that our body is designed for sporadic calorie consumption rather than continuous overfeeding, and it is true that overeating is linked to many chronic diseases.

In a 2015 experiment conducted and authored by over 20 scientists, Brandhorst and colleagues found that a four-day fasting diet was found to decrease the size of multiple organs and systems in mice followed by multisystem regeneration by stem cells after refeeding. Furthermore, bimonthly fasting cycles in mice extended longevity, lowered visceral fat, reduced cancer incidence and skin lesions, rejuvenated the immune system, and slowed bone mineral density loss. In old mice, fasting promoted the generation of new neurons in the brain's hippocampus and improved cognitive performance. The same authors also ran a small-size pilot study to see if these same benefits would be found in humans, finding that three fasting cycles decreased risk factors for aging, diabetes, cardiovascular disease, and cancer without major adverse effects. They concluded that, though more research is needed, these are promising results for the use of fasting to promote health span, the part of one's life spent in good health in contrast to using life span as a measure wherein no regard to health is given.

IF may also improve metabolic flexibility, which is the body's ability to switch between carbohydrate and fat as a fuel source. A 2012 study by Karbowska and Kochan on mice found that IF showed an increase in certain biomarkers associated with metabolic flexibility. This involved true fasting where the mice had repeated cycles of days without any food. Being able to switch efficiently and quickly between carbohydrate and fat as fuel sources, dependent upon nutrient availability and the physiological demands on the body, is considered to be optimal (Muoio 2014). The opposite—metabolic inflexibility—occurs in overfed individuals who do not easily switch between fat and carbohydrate as fuel. Metabolic inflexibility is thought to be the cause of insulin resistance, which is the cause of diabetes (Muoio 2014).

One of the mechanisms behind the observed benefits of fasting may be autophagy. Meaning self-eating in Greek, autophagy is the body's way of cleaning out damaged cells and is beneficial to our overall health. The body roots out and disintegrates damaged or dysfunctional cells, using the recycled parts for cellular repair and cleaning. Autophagy plays a housekeeping role in removing misfolded or aggregated proteins, clearing damaged cell parts, and eliminating intracellular pathogens (Glick, Barth, and Macleod 2010).

Autophagy is required for optimal anticancer surveillance by the immune system, and, in at least some cases, cancer proceeds along with a temporary inhibition of autophagy (Pietrocola et al. 2017). Autophagy is happening all the time in the body, but reduced or abnormal autophagy leads to the development of cancer, which is the uncontrolled proliferation or growth of cells (Yun and Lee 2018). In a review of the literature on IF as an anticancer treatment, Caccialanza and colleagues (2018) noted that there is some evidence that both IF (as in true food abstention) and calorie restriction both increase autophagy and inhibit tumor growth, which suggests we might not need complete abstention from food to obtain the benefits of fasting. Very importantly, however, they noted that while autophagy suppresses tumors in healthy cells, it may promote malignancy in cancer cells, a major concern for someone already diagnosed with cancer. Clearly, much more research is needed before declaring IF or calorie restriction as an anticancer treatment, and the authors advised a cautious approach.

IF may also benefit the brain. It has been well known that IF induces autophagy in the body, but it was not until 2010 that researchers found that the same happens in the brain. Alirezaei and colleagues (2010) found that short-term fasting induces profound neuronal autophagy in mice. Without appropriate autophagy in neurons, neurodegenerative disease like Alzheimer's disease can proceed. Like the other IF studies, though, the experiment by Alirezaei and colleagues was conducted in mice, and whether results obtained by fasting in cellular and animal models can be transferred to humans is still to be determined. As of now, the evidence provided by human studies is still very limited.

Fasting as Medicine?

Is it possible that one day doctors will prescribe fasting for patients for a variety of diseases? While not impossible, fasting can be a difficult thing to prescribe. One of the goals of IF research is to unlock the mechanism behind fasting's ability to boost autophagy and distill that mechanism into a usable medicine, but we are still very far from that possibility. So, the only way to experience the benefits of IF for now is to practice IF, which makes prescribing it very unorthodox. Not everyone with the early signs of cancer will react positively to the advice of eating less, and, considering that IF has even been seen to induce malignancy, much more research on IF is needed. When faced with a serious illness such as cancer, patients usually want to know that they are receiving the most state-of-the-art medicines and most effective treatments. But this might be where yoga comes in.

While a patient might be less than pleased to receive fasting advice from a doctor, the practice of yoga presents a different way of bringing fasting into a person's life. The very essence of yoga is overcoming challenges and developing self-will. Just showing up to a yoga class requires one to overcome the pull of the sofa, a daily practice of willpower. It fits that IF, then, can be a natural complement to a regular yoga and movement practice.

Is Fasting Right for Everyone?

While fasting does not seem to carry any major side effects, people who are new to IF or calorie restriction commonly experience fatigue, lethargy, dizziness, and headaches when starting such a regimen. These symptoms usually ease off, though, as the body adapts to the regimen. It should be noted that there is a fine distinction between consciously choosing to fast and skipping lunch because you are too busy at work. While fasting may have many benefits, the stress of working under pressure might counteract the benefits of consciously choosing to fast.

Fasting might not be right for everyone and should even be avoided by certain people. Women who are pregnant or breastfeeding, children, and people recovering from surgery should not try IF. Additionally, people with a history of eating disorders should completely avoid IF as anorexia can easily be masked with a veil of practicing wellness. Furthermore, if not balanced well, IF can easily lead to malnutrition and even sarcopenia, which is loss of muscle mass. Finally, people who are underweight or have osteoporosis should speak to a doctor before commencing an IF regimen.

Finally, there is another way to obtain many of the benefits of IF: exercise including yoga. Many of the benefits of IF are also seen with exercise. Exercise is known to improve metabolic flexibility (Goodpaster and Sparks 2017) and boost autophagy in the body (Drake, Wilson, and Yan 2016). Again, while there is no research directly focusing on yoga and autophagy, a lunge is a lunge, and a regular movement practice is an essential part of a healthy lifestyle.

The Yogic Diet

The *Yoga Sutras of Patanjali* and the *Bhagavad Gita* make no mention of asana other than a seated meditation position. In the same way, they are silent on what constitutes a yogic diet. Though no definition of a yogic diet exists, yoga is often associated with vegetarianism and Ayurveda. Here we explore these two concepts.

Do You Have to be a Vegetarian to be a Yogi?

Swami Sivananda, founder of Sivananda Yoga, promoted a lacto-vegetarian diet, stating that meat eaters absorb "the fear and pain of the slaughtered animal" (Sivananda Yoga Europe, n.d.). Yogi Bhajan, who is credited with bringing kundalini yoga to the West, also recommended a lacto-vegetarian diet and referred to the adage that you are what you eat, so, "Don't make your body a junkyard or a graveyard" (Kaur n.d., para. 6). Cofounder of Jivamukti Yoga, David Life, however, disagrees with incorporating dairy into the diet. Vegetarian since the 1970s and vegan since 1987, Life actively encourages yogis to be vegans in honor of *ahimsa*, or nonviolence. People attending Jivamukti teacher trainings have famously been required to watch documentaries showing the cruel practices of certain animal husbandry operations in the United States (Hagen 2019).

On the other hand, according to an article from *Yoga Journal*, Sianna Sherman, creator of Rasa Yoga, has a diet that includes fish and milk. Ana Forrest, founder of Forrest Yoga, says she eats meat, though only wild game. And finally, Scott Blossom, Ayurvedic educator and yoga teacher, occasionally eats red meat, which he considers medicine for his specific constitution, though he otherwise largely follows a vegetarian diet (Macy 2017).

Our digestion and gut flora are unique to us like a fingerprint, and our dietary needs can vary with the fluctuations of our lives. Perhaps finding the right yogic diet is about finding what works best for you including the ethics of your food choice. Only you can truly know what works best for your digestion as you are the only one to have the experience of your body in the same way that you are the only one experiencing yoga in your body, and so you must choose the yoga that fits your needs. No scriptures can tell us which that is.

Ayurvedic Diet

Present for thousands of years, Ayurveda is a traditional system of medicine, in contrast to scientific medicine. Ayurveda is based on the principle of balancing different types of energy in the body and finding balance within nature, which is believed to improve health. A central tenet of Ayurveda is that treatments, whether they are food, medicine, or other, are personalized to the individual's energy type, which separates it from other diets or systems of medicine.

According to Ayurveda, the universe is composed of five elements: *Vayu* (air), *Jala* (water), *Akash* (space), *Teja* (fire), and *Prithvi* (earth). These elements then form three different doshas, or energies, that circulate within the body and are responsible for certain physiological functions. The pitta dosha, for example, is said to control hunger, thirst, and body temperature; while the vata dosha is said to maintain electrolyte balance and movement; and finally, the kapha dosha promotes joint function (Jaiswal and Williams 2016).

An important aspect of an Ayurvedic diet is the assessment of a person's dominant dosha, then following guidelines for when, how, and what to eat based on your dominant dosha so as to promote balance among all three. The main characteristics for each dosha are:

- *Pitta (fire + water)*. Intelligent, sharp-witted, and decisive. A pitta-dominant person usually has a medium physical build and a short temper, and they may be prone to overheating. Pitta imbalances often manifest in the body as infection, inflammation, rashes, ulcers, heartburn, and fever.

- *Vata (air + space)*. Creative, energetic, and lively. A vata-dominant person usually has a thin, light frame and may be prone to digestive issues, fatigue, anxiety, dry skin, or constipation.

- *Kapha (earth + water)*. Naturally calm, grounded, and loyal. Those with a kapha dosha often have a sturdy frame and may have issues with weight gain, asthma, depression, or diabetes. Psychologically, the loving and calm disposition of kapha imbalance can transform into lethargy, attachment, and depression.

According to the Ayurvedic diet, you should eat foods that help balance out your doshas. For example, pitta-dominant people should focus on cooling, energizing foods and limit spices, nuts, and seeds. Those with vata dominance should have warm, moist, and grounding foods and restrict dried fruits, bitter herbs, and raw veggies. Finally, the kapha-dominant individual should limit heavy foods such as nuts, seeds, and oils in favor of fruits, veggies, and legumes. For all three doshas, the Ayurvedic diet recommends reducing red meat, artificial sweeteners, and processed ingredients while increasing healthy whole foods, which is good dietary advice for anyone.

Western medicine focuses on treating disease while Ayurveda focuses on treating the individual. In Western medicine, medicinal drugs are isolated from more complex sources with the aim of finding the ideal dosage for treating disease. The antibiotic penicillin, for example, was originally derived from some common molds called *Penicillium*. Ayurveda, on the other hand, applies a different approach: The prime focus is the individual rather than the disease. Both methods have their merits. As the scientific method normally looks at singular treatments or interventions while controlling other variables, applying the scientific method to individualized nature of Ayurveda proves difficult. What makes Ayurveda special—its individualized treatment—makes it impossible to apply to the general population.

Nonetheless, the Ayurvedic diet does promote some sound principles that can be easily applied to everyone: encouraging the consumption of whole foods, discouraging the consumption of processed foods, and promoting mindful eating. Following

Ayurveda, all are recommended to eat nutrient-dense whole foods such as fruits, vegetables, grains, and legumes, no matter what their dominant dosha is. Eating fiber-rich whole foods like these may help with weight loss. A few small studies have found obese people were able to lose weight following an Ayurvedic diet (Sharma et al. 2009) or an Ayurvedic-based lifestyle modification program that included dietary changes and yoga classes (Rioux, Thomson, and Howerter 2014). Perhaps that is not very surprising given that nearly all diets seem to work in losing weight. The key is to maintain that lowered weight.

The Ayurvedic diet also discourages processed foods, which often lack fiber and important vitamins and minerals. Some research has linked ultra-processed foods with heart disease, cancer (Srour et al. 2019), and even shortened life span (Schnabel et al. 2019).

Ayurveda is not just concerned with what you eat but how you eat it. Mindfulness is about paying close attention to how you feel in the present moment. Eating mindfully then involves minimizing distractions during mealtimes to instead focus on the taste, texture, and smell of the food and how that food makes you feel. A pilot study (Dalen et al. 2010) showed that eating mindfully reduces body weight, depression, stress, and binge eating as well as enhances self-control and promotes a healthy relationship with food (Kristeller and Jordan 2018). Whether you know your dominant dosha or not, we can surely all benefit from the basic principles of an Ayurvedic diet.

CONDITIONS OF THE DIGESTIVE SYSTEM

Digestive problems are very common, perhaps because our digestive system is profoundly affected by our diet and lifestyle, which can vary widely from one day to the next. Inflammatory bowel disease and irritable bowel syndrome are discussed here.

Inflammatory Bowel Disease

Inflammatory bowel disease (IBD) is a term for two conditions—Crohn's disease and ulcerative colitis—that are characterized by chronic inflammation of the gastrointestinal tract. While the exact cause of IBD is not completely understood, it is known to involve a combination of genetics, the immune system, and environmental factors. The immune system normally attacks and kills foreign pathogens, such as bacteria, viruses, fungi, and other microorganisms. With IBD, the immune system organizes an inappropriate response to the intestinal tract, resulting in inflammation. Certain environmental factors, such as food or emotional stress, trigger the harmful immune response in the intestines.

IBD is not to be confused with irritable bowel syndrome (IBS). Although people with IBS may experience some symptoms similar to IBD, the two are very different. IBS is not caused by inflammation, and the tissues of the bowel are not damaged as they are in IBD.

A significant proportion of IBD patients report or suffer from depression, anxiety, or both. Fatigue and sleep disorders are also fairly common in people with IBD, and stress is associated with a higher relapse of IBD. The prevalence of sleep disorders in IBD ranges between 44 and 66 percent as opposed to 27 to 55 percent in healthy control subjects and 67 to 73 percent in people with IBS. These complications affect social functioning and quality of life as well as being associated with anxiety or depression, disease activity, reduced physical activity, medication use, and anemia (Torres et al. 2019).

In 2019, Torres and colleagues reviewed all the literature on complementary therapies for the treatment of IBD. They noted that patients with IBD increasingly turn to alternative and complementary therapies with up to half of all of these patients using various forms of complementary and alternative medicine at some point in their course of the disease. The authors considered studies that have looked at herbal and dietary supplements, including cannabis; mind–body therapies, including yoga and meditation; and body-based interventions, including acupuncture, chiropractic, and osteopathy. Regarding yoga, the authors concluded that there is evidence that it can be effective to reduce IBD symptoms, though the evidence is limited. Moreover, they also determined that yoga improves quality of life for people with IBD, which means that their symptoms might still be present, but their overall well-being is improved.

Irritable Bowel Syndrome

Up to a fifth of Americans experience irritable bowel syndrome (IBS) symptoms, but the condition affects more women than men (Grundmann and Yoon 2010). While symptoms can be quite minor for some, they can be a major life disrupter for others. IBS is a group of gastrointestinal symptoms that usually occur together. Though symptoms can vary between people, the diagnostic for IBS is that the symptoms last at least three days per month and three months at a minimum.

While IBS can be painful, it usually does not cause intestinal damage. The symptoms include cramping, abdominal pain, bloating and gas, constipation, and diarrhea. And it is possible for someone with IBS to have episodes of both constipation and diarrhea.

The cause of IBS can vary from one person to another, but IBS is a stress-sensitive disorder (Qin et al. 2014). Women may have symptoms around the time of menstruation, or their symptoms may increase during this time, and menopausal women have fewer symptoms than women who are still menstruating. Some women have also reported that certain symptoms increase during pregnancy. While symptoms of IBS in men are the same as those in women, substantially fewer men report their symptoms and seek treatment.

In a 2016 review, Schumann and colleagues found that while yoga cannot cure IBS and thus cannot be recommended as a routine intervention, yoga can be a feasible and safe adjunctive treatment for people with IBS.

CONCLUSION

The digestive system is an intricate and complex organization of many organs and microbiota that allows us to synthesize materials from the outside world, giving us the required energy to do all the activities we do in our everyday lives. The microbiota of our gut, which number in similar quantities to our own human cells, have a profound impact on our overall health, and microbiota dysbiosis has been linked with neural diseases such as Parkinson's disease and autism. Mindful eating can help us not just to enjoy our food better but also to consider how our food choices make us feel. Research on intermittent fasting suggests that it can improve metabolic flexibility and autophagy, but it might not be right for everyone, and exercise can provide many similar benefits. There is not one definition of the yogic diet, but Ayurvedic principles provide sound guidance for a balanced diet and ask the practitioner to reflect on their individual constituency. Though yoga cannot cure inflammatory bowel disease or irritable bowel syndrome, the literature suggests that it can improve symptoms. Our digestive system affects and is affected by all other systems of the body, so looking after our digestive health is an important aspect of taking the best care of our whole being.

Based on the science presented in this chapter, the following guidelines are probably good advice for most people:

1. Take your time to eat.
2. Avoid eating when feeling emotional.
3. Eat whole foods and reduce your intake of processed foods.
4. If you are healthy and eat a balanced diet, supplements are probably not needed.
5. While alcohol may have health benefits in moderation, avoid binge drinking.
6. Continue moving and exercising. Yoga is great, of course!

Beyond following generalized advice, understanding the nuances of our own unique digestive systems can have profound benefits. Here are some questions to consider about your own digestion:

1. Do I eat enough whole foods, including fruits and vegetables?
2. Does eating spicy food affect my digestion?
3. What is the ideal time for me to eat?
4. Does eating late affect my digestion and overall well-being?
5. What are some foods that I enjoy but I know are not good for me? Could I substitute those foods for something that agrees with me better?
6. How does stress affect my digestion?
7. Do I take time out for rest and relaxation?
8. Do I take time to savor my food?
9. Do I practice yoga or exercise the right amount?

Just as each person's yoga practice is unique, so, too, is our digestive system. Through yoga, we come to know ourselves. Applying that same process of self-inquiry to what we eat and how we eat can have profound implications to our overall health.

Chapter 9

PRACTICE WITH CONFIDENCE

Throughout this book, we have explored how yoga affects the many systems of the body as far as the scientific literature shows us. We have shown that, though some of the claims made about the benefits of yoga are likely myth, yoga as a whole offers many benefits. While it is unlikely that Shoulder Stand (Salamba Sarvangasana) stimulates your thyroid gland or that twists detoxify your liver, some of the many benefits of yoga include calming the central nervous system toward the relaxation response, lowering your risk of cardiovascular disease, as well as understanding and appreciating your body with its individual needs and differences.

This chapter offers four different styles of yoga practice: a strong, dynamic practice; a slower hatha practice; a chair yoga practice; and a restorative yoga practice.* If you have ever wondered whether one style of yoga is better than another, Cramer and colleagues (2016) examined that very question in a systematic review, concluding that research studies with different yoga styles do not differ in their odds of reaching positive outcomes, and therefore the choice of an individual yoga style can be based on personal preferences and availability. Each of the practices in this chapter includes some of the physiological principles explored in earlier chapters. Though we hope that this does not need to be said, remember that you are the authority for your body, and your yoga should be modified as needed.

If you are working with an injury or have a preexisting medical condition, please get confirmation from a medical professional that you can safely participate before taking part in these practices. Move mindfully, with control, and stay within your pain-free range of movement. You always have the option to decrease the size of the movements that you are making, slow things down, or rest at any time.

STRONG, DYNAMIC PRACTICE

In this practice, we explore some of the concepts covered in the previous chapters, including tensegrity, active versus passive range of motion, and using eccentric training to build strength. Considering that yoga is all about balance, we offer this practice with the idea of leading with the left side to balance out the idea that every sequence must always start with the right side.

* You can access free videos of these practices online at www.thephysiologyofyoga.com.

We recommend reading through the whole practice while seated so that you know what to expect, then physically following the practice using the images.

For this practice, you will need a yoga brick and two straps. If you do not have these specific yoga props at home, you can get creative and make use of other household items such as a sturdy box and a trouser belt.

To prepare for the practice, make a loop in your strap and place your yoga bricks near the front of your mat.

BREATH MEDITATION

Begin in any comfortable sitting position from a Seated Cross-Legged Pose (Sukhasana; figure 9.1*a*) to Hero Pose (Virasana; figure 9.1*b*), using bricks or bolsters under your seat as needed. Notice your breathing, simply observing the inhalation and exhalation. Then, begin to lengthen each breath cycle, drawing out the inhalation and exhalation but not to the point that it feels strained. Next, add a short breath retention at the top of the inhalation, then slowly exhale, trying this for a minute or two. Notice how slowing the breath makes you feel.

FIGURE 9.1   Breath meditation: *(a)* Cross-Legged Pose; *(b)* Hero Pose.

KAPALABHATI

FIGURE 9.2   Kapalabhati.

While kapalabhati (figure 9.2) might not stop the aging process (see chapter 3 for more details), you might find it energizing. Focus on a strong, forceful exhalation and relax on the inhalation. You might imagine a small feather has landed on your upper lip and you are trying to blow it off by quickly exhaling through your nose. The abdominals should contract inward as you exhale. You can place a hand on your abdomen to feel this movement.

To begin, remain in a comfortable seated position. Close your eyes if you like, and take a few long, slow breaths. Then, take a good-sized inhalation and begin the forceful exhalations. Follow a slow tempo, around one exhale per second. After 30 cycles, take three large bellows breaths (bhastrika). Then, inhale to three-quarters and hold your breath (kumbhaka) for 30 seconds if tolerable.

Repeat the whole process for another one or two sets, gradually increasing the tempo of the exhalations and, if you like, the number of breath cycles per set.

SUN SALUTATION (SURYA NAMASKAR)

At the top of the mat, come into Mountain Pose (Tadasana)—stand tall with your feet together and hands in prayer position in front of your chest. Begin cultivating ujjayi breathing by gently constricting the vocal folds to make a soft *H* sound, which helps focus the mind on the breath and helps to maintain a steady degree of intra-abdominal pressure (see chapter 3 for more details).

Explore your use of ujjayi breathing to see when it works in your practice and when it does not, rather than thinking of it as mandatory for the whole practice. Certain movements or asanas might lend themselves better to ujjayi breathing than others. For now, focus on taking five slow ujjayi breaths in Mountain Pose before commencing your Sun Salutations.

From Mountain Pose with hands in prayer, inhale and stretch your arms up to Upward Salute (Urdhva Hastasana; figure 9.3*a*). Exhale to Forward Fold (Uttanasana; figure 9.3*b*). Inhale, reach your chest forward, and look forward in Half Forward Fold (Ardha Uttanasana; figure 9.3*c*). Exhale and step your left leg back (figure 9.3*d*). Inhale and reach your arms up for High Lunge (Ashta Chandrasana; figure 9.3*e*), with the palms of your hands together. Exhale and bring your hands down to frame your front foot, then inhale into High Plank (Phalakasana; figure 9.3*f*). Exhale and lower to Low Plank (Chaturanga Dandasana; figure 9.3*g*) with your elbows close to your ribs. If needed, place your knees on the floor as you lower. Push forward with your toes and inhale into Upward-Facing Dog (Urdhva Mukha Svanasana; figure 9.3*h*), but feel free to instead come into High Cobra (Bhujangasana) or even Low Cobra with little or no weight on your hands. Exhale to Downward-Facing Dog (Adho Mukha Svanasana; figure 9.3*i*). Remain here for three slow breaths. Then, inhale and lift your left leg to Downward Dog Splits (Eka Pada Adho Mukha Svanasa; figure 9.3*j*). Exhale and step your left foot between your hands (figure 9.3*k*). Inhale and step your right foot forward and lift to Half Forward Fold (figure 9.3). Exhale and soften into Forward Fold (figure 9.3*m*). Inhale, reach your arms up (figure 9.3*n*), and finally exhale to bring your hands into prayer position for Mountain Pose (figure 9.3*o*).

Then repeat the entire sequence, leading with your right leg. You will then have completed one full round. Do two more rounds on each side, completing a total of three full rounds.

FIGURE 9.3   Sun Salutation: *(a)* Upward Salute; *(b)* Forward Fold; *(c)* Half Forward Fold; *(d)* left foot steps back; *(e)* High Lunge; *(f)* Plank; *(g)* Low Plank; *(h)* Upward-Facing Dog;

(i) Downward-Facing Dog; *(j)* Downward Dog Splits; *(k)* left foot steps forward; *(l)* Half Forward Fold; *(m)* Forward Fold; *(n)* Upward Salute; *(o)* Mountain Pose.

CHAIR POSE (UTKATASANA) AND VINYASA

FIGURE 9.4   Chair Pose.

Begin in Mountain Pose with your feet together. Then, lift your arms overhead as you bend your knees, as if to sit in an invisible chair while keeping your weight on your heels (figure 9.4). Notice the oppositional energies needed to hold this pose. Reaching your arms up makes your front ribs want to flare forward so you engage your abdominals at the same time. Try to enjoy these dueling energies as your thighs work hard to keep you steady. It is little wonder that *utkata* means fierce or awkward. You might feel both fierce and awkward as you hold this pose for five breaths, then fold forward and flow through a vinyasa, ending up in Downward-Facing Dog.

REVOLVED HIGH LUNGE SEQUENCE

FIGURE 9.5   Revolved High Lunge.

Standing poses are great for building heat and strength in the body. At this point of our practice, we explore a Revolved High Lunge sequence and a wide-legged sequence.

Revolved High Lunge offers an opportunity to explore the idea of mobility and flexibility or active and passive ranges of motion to see how one might affect the other. Throughout the High Lunge (figure 9.5), try to keep your pelvis facing forward and not rotating with your spine so that we can purely explore spinal rotation.

ACTIVE REVOLVED HIGH LUNGE

FIGURE 9.6   Active Revolved High Lunge.

From Downward-Facing Dog, step your left foot forward and come into High Lunge. As you exhale, rotate your torso to the left with arms parallel to the floor (figure 9.6). Note how far you can twist here—a twist that comes from the contraction of the trunk muscles—and note how much those muscles must engage. You are not leveraging against anything, so this spinal rotation could be considered your active range of motion, or what some people refer to as *mobility*.

ASSISTED REVOLVED HIGH LUNGE

Place your right hand outside your left knee and leverage against your hand to deepen the spinal rotation (figure 9.7). This external rotational force created by your hand against your knee allows you to explore your passive range of motion, or flexibility, in spinal rotation.

FIGURE 9.7   Assisted Revolved High Lunge.

ACTIVE REVOLVED HIGH LUNGE (AGAIN)

Return your right arm to horizontal, noticing whether your active mobility has been affected by your exploration of the small degree of passive spinal rotation available. Finally, release out of the pose and step back to Downward-Facing Dog.

Optional Vinyasa and Second Side

If a Vinyasa is calling to you, then go ahead and flow through. Then, after a few breaths in Downward-Facing Dog, repeat the whole lunge sequence on the second side.

WIDE-LEGGED FORWARD BEND (PRASARITA PADOTTANASANA)

With the next two poses, we explore the difference between coming to the end of our range and holding back slightly.

From Downward-Facing Dog (Adho Mukha Svanasana), float your left leg up behind you, then lightly step your left foot between your hands. Come up to stand as you turn to the long edge of the mat, preparing for Wide-Legged Forward Bend. With your feet parallel to each other and toes

FIGURE 9.8   Wide-Legged Forward Bend.

pointing to the long edge of the mat, interlace your fingers behind you and fold forward (figure 9.8). Go to your edge in this stretch and stay there for five breaths. Then, come up about an inch so that you are aiming for 90 percent of your maximal stretch and stay here for five breaths. When you pull back from your end range of motion, the muscles that are being stretched must engage to help stabilize you (a practice we recommend in chapter 1 for people with hypermobility). Note how the pose feels when you simply pull back from your end range a little bit.

SIDE LUNGE (SKANDASANA)

a

b

FIGURE 9.9   Side Lunge: *(a)* 80 percent of full flexion; *(b)* full flexion.

Come all the way up from Wide-Legged Forward Bend (Prasarita Padottanasana). Then, lower yourself to Side Lunge (Skandasana; figure 9.9*a*) on your left but hold at about 80 percent of your maximum knee bend. Staying here requires engagement of many muscles around the thighs. Hold for five breaths.

Lower yourself so that your buttock rests on your left heel or as near as you can (figure 9.9*b*). This approach to the pose focuses on passive flexibility. After staying for five breaths, try to come up from here. Notice how the concentric phase (coming up) is much harder to do than the eccentric phase (lowering). Repeatedly lowering to the floor with control (then using your hands on the floor to assist you in coming up) can be a good way of building strength so that one day you can also come up easily. Finally, step back to Downward-Facing Dog (Adho Mukha Svanasana).

 Optional Vinyasa and Second Side

If a vinyasa is calling to you, then go ahead and flow through. Then, after a few breaths in Downward-Facing Dog (Adho Mukha Svanasana), repeat the wide-legged whole sequence on the second side.

DANCER'S POSE (NATARAJASANA)

FIGURE 9.10   Dancer's Pose.

Balancing poses offer an opportunity to focus the mind, and good balance is essential in preventing falls. Here, we explore two versions of Dancer's Pose (Natarajasana).

From Downward-Facing Dog, jump or walk to the front of the mat. Then, come all the way up and finally settle in Mountain Pose (Tadasana). Press down evenly through your feet and focus your drishti ahead. Come into Dancer's Pose by taking hold of the outside (or, if you prefer, the inside) of your left foot. Kick back through your left foot as you reach your chest forward and up. Let your left leg pull your arm back (figure 9.10) and hold for five low breaths, then return to Mountain Pose (Tadasana).

KING DANCER'S POSE (VARIATION OF NATARAJASANA)

FIGURE 9.11 King Dancer's Pose (with strap).

Though many poses have elements of tensegrity, this variation in particular feels like an embodiment of the idea of tensegrity, or tensional integrity (as discussed in chapter 1).

Except for the very few who are superflexible, you will need a strap to bind in this version of Dancer's Pose (Natarajasana; figure 9.11). Loop the strap around your left foot and kick your foot back as you take hold of the strap with both hands overhead. Walk your hands down the strap toward your foot, elbows pointing upward and lifting your left foot high. Notice the elements of tension and compression working together to bring stability to the pose. Hold for 5 (or if you can, 10) breaths, then release from the pose with control and come to Mountain Pose (Tadasana).

 Second Side of Both Poses

Do both poses on the second side (your right foot). Then, do one Sun Salutation but finish in Downward-Facing Dog (Adho Mukha Svanasana) or skip the Sun Salutation and just step back to Downward-Facing Dog (Adho Mukha Svanasana).

BOW POSE (DHANURASANA)

FIGURE 9.12 Bow Pose.

Strengthening of the spinal extensors is associated with a number of benefits. This sequence can help to mobilize and strengthen this important muscle group.

From Downward-Facing Dog (Adho Mukha Svanasana), transition to a prone position, that is, lying on your front. Next, bend your knees and take hold of your ankles or feet. As you inhale, kick your feet back, lifting your chest, knees, and thighs off the floor (figure 9.12). See if you can keep your knees hip-width apart and point your toes. Look forward or wherever is comfortable for the neck and hold the pose for five breaths. Then, release slowly, letting go of your feet. You can make a pillow with your hands and rest your head to the left for a few seconds.

Repeat the pose a second time, turning your head to the right afterward to rest.

LOCUST (SALABHASANA)

FIGURE 9.13   Locust Pose.

Still lying on your abdomen with your legs straight behind you and the tops of your feet on the floor, with your arms alongside your body and your palms facing the floor, rest your forehead lightly on the mat. As you inhale, lift your head, chest, and arms off the floor, reach through your fingers while pressing the tops of your feet into the floor and softly focusing your eyes a few feet in front of you (figure 9.13). Hold for five breaths, then rest by turning your head to the left. Make a pillow with your hands if you like and stay here a few moments.

Repeat the pose a second time, turning your head to the right afterward to rest, and, finally, come back to Downward-Facing Dog (Adho Mukha Svanasana).

BOAT POSE (PARIPURNA NAVASANA)

FIGURE 9.14   Boat Pose.

Though the spine is not in extension in Boat Pose, the spinal extensors still must work to maintain a neutral spine. From Downward Dog, jump or walk forward to come to a seated position with your legs bent and feet on the floor in front of you. As you inhale, come into Boat Pose (figure 9.14). Extend your knees as you lift your chest and draw the shoulders back and down. At the same time, stretch your arms forward so they are parallel to the floor. Hold for five breaths. To come out, see if you can keep your legs where they are and slowly roll your torso down to the floor.

SHOULDER BRIDGE (SETU BANDHASANA)

While you are on your back, bend your knees and walk your feet close to your

FIGURE 9.15   Shoulder Bridge.

seat, keeping them roughly hip-width apart. Brings your arms alongside your body with your palms facing down and your fingers slightly touching your heels. Inhale, press into your feet and shoulder blades and lift your hips up. Interlace your fingers underneath you and try to walk your shoulders under your ribs (figure 9.15). Strongly press down into your arms, shoulders, and hands. Hold for five breaths, then slowly lower yourself to the floor.

WHEEL POSE (URDHVA DHANURASANA)

FIGURE 9.16   Wheel Pose.

Lie on your back. Bend your knees and walk your feet close your seat, keeping them roughly hip-width apart. Bend your elbows, placing the palms of your hands beside your head with your fingertips pointing toward your shoulders. See if you can keep your feet parallel and knees in line with your feet. Inhale and press your hands and feet firmly into the floor as you lift your hips, shoulders, and head off the floor (figure 9.16). Hold for five breaths if you can, then exhale and tuck your chin into your chest, lowering with control.

If you have it in you, repeat Wheel Pose (Urdhva Dhanurasana) once more, then hug your knees to your chest and rock side to side.

LIZARD OR PIGEON (EKA PADA RAJAKAPOTASANA)

FIGURE 9.17   Lizard Pose.

Now that we have reached the peak of the class, it is time to enjoy the downward slope of the practice as we settle into some long-held deep stretches before the final relaxation.

From Downward Dog, inhale and raise your right leg. Exhale and place your right foot on the outside of your right hand with your right leg externally rotated so your toes point outward. Lower your left knee to the floor and bring your forearms to the floor or onto bricks (figure 9.17). Hold for 10 slow breaths.

PIGEON (EKA PADA RAJAKAPOTASANA)

FIGURE 9.18   Pigeon Pose.

From Lizard, simply wiggle your right foot to the left edge of the mat until your right ankle is behind your left wrist. Inhale and lift your chest. As you exhale, fold from your right hip crease to lower your torso (figure 9.18). Consider using your brick or hands as a head rest. Stay here for 10 breaths. Then, press back to Downward Dog.

Repeat Lizard and Pigeon on the left side.

FINAL RELAXATION: CORPSE POSE (SAVASANA)

FIGURE 9.19 Corpse Pose.

The most common Savasana position is lying on your back with your arms and legs extended (figure 9.19). However, this is not the only option for your final relaxation. You could, for example, have your knees bent with the soles of your feet on the floor and your hands resting on your tummy. You could rest your feet on a nearby sofa or chair, or simply slide a bolster under your knees. The most important thing is that you are in a position in which you can fully relax.

Take a few long, slow breaths through your nose. Notice the points of your body in contact with the floor. Let yourself relax further into those points of contact, softening into the floor. Let all tension melt away and give yourself permission to completely surrender, completely relax. Stay in Corpse Pose (Savasana) for 5 to 10 minutes or longer if desired.

Slowly come out of Savasana. You might slowly wiggle your fingers and toes, then reach your arms back for a long, full-body stretch. Slowly ease yourself up to a comfortable sitting position. Take a moment to reflect on how you feel after the practice. Observe your breath for a few moments. Now bring your hands into prayer position at the front of your chest. Take a deep inhale, and as you exhale, bow your head toward your hands in honor of the yoga practice.

SLOW HATHA PRACTICE

Though *hatha* is popularly explained as the uniting of sun and moon, this is a more modern interpretation of the phrase (Birch 2011). In Sanskrit, *hatha* means force, and the term *hatha yoga* was more likely inspired by this meaning (Birch 2011; Mallinson 2011). While modern yogis might not associate yoga with force, this likely refers to the effort required to create its effects (Mallinson 2011) or to the forceful moving of subtle esoteric energies known as *kundalini*, *apana*, and *bindu* upward through the central channel (Birch 2011). Though metaphysical energetics are beyond the scope of this book, we authors have found that newcomers to yoga often comment on how the practice is much more challenging or harder than the 60 minutes of relaxation that they were expecting. Indeed, more force was required than they expected.

This practice might challenge you, and it will certainly require effort, but you can adjust the practice to your own level by reducing the intensity with which you practice, modifying poses, or simply resting as needed. As opposed to more dynamic practices, hatha is all about staying in poses slightly longer to explore them more deeply and to build strength through isometric holds. This practice leads with the right leg, as is commonly guided.

RECLINING BREATH MEDITATION

FIGURE 9.20 Reclining breath meditation.

Lie on your back with your knees bent and the soles of your feet on the floor. Let your knees roll in and rest against each other. Then, rest your hands on your belly and rib cage (figure 9.20). Take a big breath in and hold, noticing how your belly and ribs expand outward. Then exhale very slowly, observing the effortless softening of your torso. Visualize the breath in a three-dimensional way. Visualize your lungs expanding at the front and back, top and bottom, left and right. Continue observing the breath and enjoy these precious moments of letting the mind settle on one thing. Nobody requires your attention right now. Nothing needs to be done right now. Enjoy this moment of simply being.

Once you feel relaxed and present, begin to slowly move your body, eventually stretching your arms back and legs forward like you have just woken up.

RECLINING HAND-TO-TOE POSE A (SUPTA PADANGUSTHASANA A)

FIGURE 9.21 Reclining Hand-to-Toe Pose A.

Reclining poses are a great way to enter a practice slowly, particularly if starting with Sun Salutations sounds less than appealing. Supine poses can also be a good way for someone with low back pain to start a practice.

Loop the center of your yoga strap over the arch of your right foot. Hold one end of the strap in each hand and slowly extend your knee until you feel a stretch on the hamstrings (figure 9.21). Your knee might not fully extend, and you might even prefer to maintain a slight bend in your knee. What matters is that you feel the sensation of stretch on the back of your thigh. Gently press the heel of your right foot away from you as you draw the top of your right foot back toward you. You have the option here to keep your left knee bent with your foot flat on the floor or to straighten your left leg out in front of you on the mat. Take a few long, slow breaths before trying a stretching technique called *proprioceptive neuromuscular facilitation stretching* described in chapter 1 (see page 22).

While holding the stretch, engage your hamstrings without moving (an isometric contraction) by pushing gently against the stretch without actually moving for 10 seconds. Then, relax into the passive stretch again for 10 seconds. Repeat this pattern twice more.

RECLINING HAND TO TOE POSE B (SUPTA PADANGUSTHASANA B)

FIGURE 9.22 Reclining Hand-to-Toe Pose B.

Take hold of both ends of the strap with your right hand and slowly lower your right leg to the right side while keeping your left buttock somewhat grounded (figure 9.22). At the limit of your movement, pause for five slow breaths, noticing where you feel the stretch. Finally, bring your right leg back to the center.

RECLINING HAND-TO-TOE POSE C (SUPTA PADANGUSTHASANA C)

FIGURE 9.23 Reclining Hand-to-Toe Pose C.

Let your right leg cross all the way to the left of the mat until you come into a spinal twist (figure 9.23). Shuffle your hips a few inches back to the center of the mat so that your outer left hip is on the midline of the mat. Feel free to place a brick under your right leg for support. Stay there for five breaths, being aware of each inhale and exhale as well as the different sensations in your body in this asana. Finally, bring your leg back up to center.

Second Side

Slowly lower your right leg to the mat and stretch your legs out long on the mat in front of you as you stretch your arms back and notice how your legs are feeling. When you are ready, repeat the sequence with the opposite side.

CAT-COW (MARJARYASANA-BITILASANA)

a

b

FIGURE 9.24 Cat-Cow: *(a)* Cat Pose; *(b)* Cow Pose.

Before moving into the Sun Salutation, we use small spinal movements to prepare for the bigger movements of the Sun Salutation.

From a supine position, slowly bring yourself onto all fours. Slowly arch and round your back. Imagine a cat rounding its back upward toward the ceiling (flexion; figure 9.24*a*) then a cow's posture where your back dips below your shoulders and hips (extension; figure 9.24*b*). Start small and build up to finding larger and larger spinal movements. Notice which parts of your spine seem to move more easily than others. Remember that your neck is part of your spine, and let it move with the rest of your spine.

DOWNWARD-FACING DOG (ADHO MUKHA SVANASANA)

FIGURE 9.25 Downward-Facing Dog.

From all fours, tuck your toes under, lift your hips, and come into Downward-Facing Dog (figure 9.25). You may need to wiggle your feet back a couple of inches to find a more comfortable Downward Dog. If your back is rounded in this pose, bend your knees a bit and see if you can hold a neutral spine. Feel free to pedal your feet as if you are walking stationary or wiggle your hips side to side to get a bit more warmed up. Then, hold still and take a few slow breaths. Walk your feet to the front of the mat and slowly come up to stand.

SUN SALUTATION (SURYA NAMASKAR)

From Mountain Pose, bring your hands into prayer position in front of your chest. As you inhale, reach your arms up, touching your palms together (figure 9.26a). As you exhale, gently fold forward over your legs and reach your hands to the floor into Standing Forward Bend (Uttanasana; figure 9.26b). With your next inhalation, extend your spine forward and look forward into Standing Half Forward Bend (Ardha Uttanasana; figure 9.26c). As you exhale, step your right leg back, placing your right knee on the floor. Inhale and reach your arms up for Low Lunge (Anjaneyasana; figure 9.26d). Exhale, bring your hands down to frame the front foot, then step your left leg back for High Plank (figure 9.26e). Exhale, place the Knees-Chest-Chin to the floor (Ashtanga Namaskar; figure 9.26f), and inhale into Cobra Pose (Bhujangasana; figure 9.26g). Exhale and come into Downward-Facing Dog (Adho Mukha Svanasana; figure 9.26h). Hold for three breaths. Then, inhale to lift your right leg up, and exhale to step your right foot between your hands. As you inhale, step your left foot forward and halfway lift, looking forward and extending the spine (figure 9.26i). Exhale to fold forward (figure 9.26j). Inhale, come all the way up, and stretch your arms up (figure 9.26k). Exhale to bring your hands in a prayer position for Mountain Pose (figure 9.26l).

Repeat on the left side. Then do two more rounds on each side.

FIGURE 9.26 Sun Salutation: *(a)* arms reach up; *(b)* Forward Fold; *(c)* spine extends forward; *(d)* Low Lunge; *(e)* Plank; *(f)* Knees-Chest-Chin; *(g)* Cobra; *(h)* Downward-Facing Dog; *(i)* spine extends forward; *(j)* Forward Fold; *(k)* arms reach up; *(l)* prayer position.

TRIANGLE (UTTHITA TRIKONASANA)

FIGURE 9.27 Triangle Pose.

Now that you are warmed up from Sun Salutations, we explore the difference between Triangle and Revolved Triangle.

From Mountain Pose (Tadasana), take a big step back with your left foot and turn it out to about 45 degrees and your right foot pointing to the front of the mat, your hips and chest facing the long edge of the mat. Inhale and reach your arms wide so they are parallel to the floor. Exhale and glide your upper body forward toward the front of the mat and place your right hand lightly on your right shin or onto a brick as you stretch your left arm up (figure 9.27). Look up toward your left hand and stack your left shoulder over your right. Once you are steady, hold for five breaths. Then, inhale and come up.

REVOLVED TRIANGLE (PARIVRTTA TRIKONASANA)

FIGURE 9.28 Revolved Triangle Pose.

Rotate your hips to face the front of the mat, pivoting the back foot as needed but still with your left foot halfway back along the mat. Inhale, place your right hand on your hip and reach your left arm up. Exhale and fold forward from your right hip crease, placing your hand inside your right foot on the floor or a brick (figure 9.28). Then, lift your right arm up until it is vertical or as close to vertical as you comfortably can lift it. Try to keep your hips squared off, meaning that your two front hip points are both level and pointing to the mat beneath you. Press firmly into both feet, stacking your right shoulder over your left shoulder. Hold for five breaths. Inhale, come all the way up, and step the back foot forward to Mountain Pose.

 Second Side

Go through the whole Triangle sequence on the other side.

HALF-MOON (ARDHA CHANDRASANA)

FIGURE 9.29 Half-Moon Pose.

Similar to the Triangle sequence, we now explore Half-Moon and Revolved Half-Moon.

From Mountain Pose, bring your right hand down to a brick as you float your left arm and left leg as high as you can, attempting to stack your left hip over your right hip (figure 9.29). Press your right foot firmly into the floor as you stretch through your left foot, as if you were trying to push a heavy object away from you with your left leg. Look upward if you can and think of emanating energy from your center and out to every direction, like a star. Hold for five breaths, then go straight into Revolved Half-Moon.

REVOLVED HALF-MOON (PARIVRTTA ARDHA CHANDRASANA)

FIGURE 9.30 Revolved Half-Moon Pose.

From Half-Moon, lower your left hand onto a brick. As you inhale, rotate to your right and float your right arm upward (figure 9.30). Look up toward your right hand if you can. As with revolved triangle, keep your hips squared off, meaning that the two front hip points are both level and pointing to the mat beneath you. Hold for five breaths. Then, place your left foot next to your right and take a well-deserved rest in a Forward Fold.

Second Side

Go through the whole Half-Moon sequence on the other side.

DOWNWARD-FACING DOG (ADHO MUKHA SVANASANA) AND COBRA VINYASA

Now we go through part of a Sun Salutation to get back to Downward Dog. From Mountain Pose, inhale and reach your arms up, touching your palms together. Exhale and fold forward over your legs to Forward Bend (Uttanasana). Inhale, extend your spine forward, and look forward into Standing Half Forward Bend (Ardha Uttanasana). Exhale and step back to Downward Dog.

COBRA (BHUJANGASANA)

FIGURE 9.31 Cobra Pose.

Backbends can be very energizing and are an important element of yoga.

From Downward Dog, come forward to lie down on your front with your legs stretched behind you and the tops of your feet on the floor. As you inhale, push into your hands and lift up to Cobra (Bhujangasana; figure 9.31), keeping a slight bend in your elbows. Press the tops of your feet and thighs firmly into the floor as you reach your heart forward. Bring your gaze upward or forward based on your preference. Stay there for five breaths and then come down and rest for a few moments.

When you are ready, come into a second round of Cobra for five breaths then rest again for a few moments.

CAMEL POSE (USTRASANA)

FIGURE 9.32 Camel Pose.

Press yourself up and come to a kneeling position with your hips over your knees. Place your hands on your sacrum or lower back and extend tall through the spine. Exhale and reach your hands back to your heels one at a time (figure 9.32). If you cannot reach your heels with the tops of your feet on the floor, tuck your toes under. If you cannot take hold of both heels, just reach your right arm back for now. You can keep your neck in line with your spine or let your head release all the way back with control as long as that is comfortable for your neck. Try connecting your tongue to the upper hard palate of your mouth. See if you can shift your hips forward a bit more, puff the chest, and then hold for five breaths. Use your core to come up, then rest on your heels in Hero Pose (Virasana) or come into Child's Pose (Balasana).

When you are ready, come into a second round of Camel and rest again for a few moments.

YOGIC SQUAT (MALASANA)

FIGURE 9.33 Yogic Squat; also known as *Garland Pose*.

The following three poses bring us step-by-step to a hand balance in Crane.

From Hero Pose, transition to Yogic Squat (Malasana). If your heels do not touch the floor, consider using yoga bricks or a folded blanket under your heels. Bring your hands to prayer position in front of your heart with your upper arms inside your knees (figure 9.33). Imagine lifting your rib cage up away from your pelvis. Focus your eyes on one point in front of you and remain steady for five breaths.

STANDING FORWARD FOLD (UTTANASANA)

FIGURE 9.34   Standing Forward Fold.

From your squat position, place your hands on the mat, straighten your legs, and bring your feet parallel as you come into Standing Forward Fold (Uttanasana; figure 9.34). Shift your weight toward your metatarsals (the balls of the feet) so that your hips are over your ankles. Let your neck relax so your head hangs heavy and hold for five breaths.

CRANE (BAKASANA)

FIGURE 9.35   Crane Pose.

From Standing Forward Fold, move your hands forward a little and bend your knees. Place your palms firmly on the floor and spread your fingers wide, grabbing the floor with your fingertips as if you were grabbing a mound of earth. Lift your heels and place your knees just below your underarms. Look forward, shift forward to your tiptoes, then lift one foot off the floor and then the other one, shifting all your weight onto your hands (figure 9.35). Hug your feet toward your seat and keep looking forward. Keep pressing down and focusing on lifting your center.

While this looks like an impressive pose, there is still an element of bone stacking, which means that the pose does not require immense muscular strength. To balance, you must shift your center of mass forward until there is no weight on your feet. Most people who fail to balance in this pose simply do not shift far enough forward. Many people feel nervous as they shift all their weight forward onto their hands. If that is the case, consider placing a bolster or cushion in front of your face to act as a crash pad just in case. When people finally get the balance right after trying for a long time, they often cannot help but applaud in excitement! Celebrate the small victories. Maybe you balance a quarter of a second, then half a second. Keep practicing, and those balance times will increase.

Finally, step back to Downward Dog.

HALF-SPLITS (ARDHA HANUMANASANA)

FIGURE 9.36   Half-Splits Pose.

We now come to the floor for a couple of poses before the final relaxation.

From Downward-Facing Dog (Adho Mukha Svanasana), inhale as you lift your right leg, and as you exhale, step your right leg between your hands. Flex your right foot and slide your right leg farther forward toward the top of the mat (figure 9.36). Place your left knee down as you inhale to look forward, maintaining a flat back for the first five breaths. For the last five breaths, let yourself soften into the pose, letting your back naturally round. To exit, step your right foot back to Downward-Facing Dog (Adho Mukha Svanasana). Repeat on the other side.

SEATED SPINAL TWIST (ARDHA MATSYENDRASANA)

FIGURE 9.37   Seated Spinal Twist.

From Downward-Facing Dog, come into a seated position with the legs in front of you and place the right foot outside the left knee. Optionally, bend the left knee so your left foot is outside your right buttock. Hook your right arm around your left bent knee and square your hips so they remain even as you come into Seated Spinal Twist (Ardha Matsyendrasana; figure 9.37). As you inhale, reach up through the crown of your head (though you are not really lengthening your spine, it will feel like you are growing taller). As you exhale, see if you can rotate a bit farther. Hold for 5 to 10 breaths. Extend your legs forward, bend your right knee and place your right ankle outside your left knee. Finally, hook your left arm around your right bent knee and twist farther. Hold for 5 to 10 breaths, release slowly, and come into a lying position on your back.

FINAL RELAXATION (SAVASANA)

FIGURE 9.38 Corpse Pose (with knees bent).

The most common Savasana position is lying on your back with your arms and legs extended. However, this is not the only option for your final relaxation. You could, for example, have your knees bent with the soles of your feet on the floor and your hands resting on your tummy. You could rest your feet on a nearby sofa or chair or simply slide a bolster under your knees. The most important thing is that you are in a position in which you can fully relax.

Take a few long, slow breaths through your nose. Notice the points of your body in contact with the floor. Let yourself relax further into those points of contact, softening into the floor. Let all tension melt away and give yourself permission to completely surrender and completely relax. Stay in Corpse Pose (Savasana; figure 9.38) for 5 to 10 minutes or longer if desired.

Slowly come out of Savasana. You might slowly wiggle your fingers and toes, then reach your arms back for a long, full-body stretch. Slowly ease yourself up to a comfortable sitting position. Take a moment to reflect on how you feel after the practice. Observe your breath for a few moments. Now bring your hands into prayer position at the front of your chest. Take a deep inhale, and as you exhale, bow your head toward your hands in honor of the yoga practice.

CHAIR YOGA PRACTICE

Chair Yoga is an accessible style of yoga that is not only great for students who find it challenging to move from a chair to the floor, and vice versa, but can easily be practiced in an office environment and even on a long-haul flight. Accessible isn't synonymous with easy: Chair Yoga can still be challenging and offers us a great opportunity to work on mobility, strength, and balance. Remember that regardless of the style of yoga, similar benefits can be had.

For this practice you will need a yoga mat, a sturdy chair (preferably without arm rests), a yoga strap, a blanket or towel, a bolster or cushion, and two foam bricks. You can always get creative with props and use household items as substitutions. Placing the chair onto your yoga mat prevents the chair from sliding on the floor while you move.

SEATED CONTEMPLATION

FIGURE 9.39 Seated Contemplation.

Place your feet roughly hip-width apart and under your knees. You always have the option to step your feet slightly farther forward at any time if you feel like you need additional support. Sit on a blanket or thin cushion if that makes you feel more comfortable, and you can use the back support of the chair if it has one. You can gently wrap a thin blanket over the back of your neck, cross it over your chest and tuck your elbows into the sides of your waist to secure the ends in place (figure 9.39). This will add some support to your head and neck.

Once you are comfortable, close your eyes or simply soften your gaze and look down slightly. Release your jaw and let your lips and teeth gently part. Allow your tongue to drop away from the roof of your mouth. Take a moment to scan your body with your mind, observing how you are feeling physically today. Try your best not to judge the sensations that you notice but simply observe. Observe your breath for a few moments, noticing its quality. Is its rate slow or fast? Is it shallow or deep? Which parts of your body are moving as you inhale and exhale? Now focus on gently inhaling and exhaling through your nose. Once you feel settled, blink your eyes open if you had closed them and remove the blanket from around your neck.

NECK MOBILITY

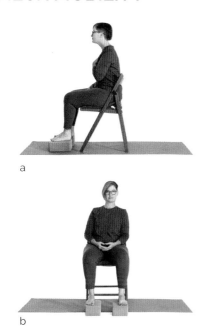

a

b

FIGURE 9.40   Neck mobility: *(a)* head moving forward; and *(b)* head moving to the side.

Slowly lower your chin toward your chest, then gently lift your chin up. Repeat this slowly for a few cycles, perhaps five times. Slowly turn your head to the right and then to the left for about five repetitions. Gently lower one ear toward your shoulder on that side. Move your head with control back to the center and then repeat on the opposite side. Repeat this for a few cycles, noticing how one side feels compared to the other. You now have the option to combine all these movements by slowly moving your head around in the circle. After a few repetitions, repeat in the opposite direction.

Place one hand on your breastbone to stabilize your thoracic spine so that it is easier to isolate the following movements of your neck. Slowly move your head forward (figure 9.40*a*) and back in the horizontal plane, without lowering or lifting your chin. After a few repetitions, return to a neutral position. Now place your index fingers directly in front of your eyes. Without rotating your head side to side, slowly move your head to the right and then to the left in the horizontal plane. You can also do this while keeping your hands in your lap (figure 9.40*b*). After a few repetitions, rest in a neutral position. Placing your index fingers in front of your eyes helps you to stabilize your upper body while you focus on isolating just your neck movements. These movements may feel a little strange and unnatural at first but with practice they will quickly become smoother and more controlled. Now you can combine all four of these movements: Place your hand back onto your breastbone; slowly shift your head forward, then to the right, to the back, and to the left; repeat this elliptical motion for a few cycles; then repeat this in the opposite direction; rest and take a few moments to notice how your neck is feeling.

SPINAL MOBILITY

FIGURE 9.41 Spinal mobility: *(a)* Seated Cat Pose with arm movements; *(b)* Seated Cow Pose with arm movements; *(c)* Seated side bend.

Placing your hands on your thighs, slowly lift through the back of your chest into a gentle backbend. Then, gently curl your spine in the opposite direction, drawing your navel back toward your spine and rounding your back. You have the option here to include coordinated arm movements; clasping your hands and reaching your palms forward as you flex your spine forward (figure 9.41*a*) and reaching your arms up and out as you extend your spine back (figure 9.41*b*). After a few rounds of this seated Cat-Cow (Marjaryasana-Bitilasana), rest in a neutral position and take a moment to notice how your spine feels.

Interlace your fingers behind your head or place your hands on your shoulders and slowly side bend in each direction, reaching up with your arm pit and elbow (figure 9.41*c*). Rest after a few repetitions and slowly twist your spine to the right and then to the left for a few repetitions, leading the movements with your rib cage.

Focusing on the lumbar spine, slowly rock your pelvis back and forward. Then, add to these movements by rolling your pelvis forward, to the right, to the back, and to left. After a few repetitions, change direction. Finally, hold still in a neutral position and observe how your lower back feels.

SHOULDER ROLLS

FIGURE 9.42 Shoulder rolls.

Resting your arms by your sides, slowly roll back one shoulder at a time (figure 9.42), first in one direction and then in the opposite direction. Then roll both shoulders back together. After a few repetitions, roll both shoulders forward together.

ARM CIRCLES

FIGURE 9.43 Arm circles.

Slowly make circles with your left arm in one direction (figure 9.43), then change direction. Gradually make the circles larger if you can. You have the option to wrap your right hand under your left underarm, allowing you to isolate the movements a little more at the shoulder joint.

Now, repeat this with your right arm, noticing how this side feels compared to the left. You also have the option to move both arms at the same time.

SHOULDER GIRDLE MOBILITY

Sitting tall, gently tense your abdomen and reach your arms forward while maintaining the position of your spine. You can also squeeze a foam brick between your hands. Slowly lift your shoulder blades up toward your ears (a movement known as *elevation*; figure 9.44a). Keep your shoulder blades elevated as you draw them together toward your spine (known as *retraction*; figure 9.44b).

a

b

c

d

FIGURE 9.44 Shoulder girdle mobility: *(a)* elevation; *(b)* retraction; *(c)* depression; *(d)* protraction.

Keep this position of retraction as you slowly draw your shoulder blades down away from your ears (known as *depression*; figure 9.44c). Keeping your shoulder blades depressed, slowly draw them apart away from your spine (known as *protraction*; figure 9.44*d*). Rest for a moment with your arms relaxed, then repeat in a new sequence: retracting, then elevating, then protracting, and finally depressing your shoulder blades. On the next cycle you can start to add some internal resistance by imagining that there is a force trying to prevent you from moving in each direction. You can also hold something slightly heavier between your hands. Notice the movements that feel particularly challenging or particularly stiff or tight.

WRIST MOBILITY

FIGURE 9.45   Wrist mobility.

Make a fist and move your hand in slow, controlled circles in one direction (figure 9.45) and then the other. Notice how much the bones in your forearm are also moving to facilitate these movements. Repeat these movements, holding onto your forearm so that the wrist joint is isolated. Finally, challenge yourself by releasing your forearm and seeing if you can move your fist without moving your forearm bones. Repeat this with your other hand and notice the difference between your right and left side.

SEATED CALF RAISES

FIGURE 9.46   Seated calf raises.

With your feet roughly underneath your knees, slowly lift and then lower your heels (figure 9.46), keeping your toes relaxed and in contact with the mat or the floor. Repeat this for a few cycles and then rest. You now have the option to add some body weight resistance by leaning forward onto your thighs and repeating these movements.

HIP CIRCLES

FIGURE 9.47   Hip circles.

Shifting your weight slightly into your left leg, lift your right foot and begin to make circular movements from your right hip joint (figure 9.47). You can choose to assist or resist these movements with your right hand. Begin to bend and straighten your knee as you move. Rest after five cycles and then move in the opposite direction. Repeat these movements on your left side.

INTERNAL HIP ROTATION

FIGURE 9.48   Internal hip rotation.

Sit slightly farther forward in your chair but make sure that you still feel supported and grounded. You can place your hands on your hips or reach your arms out in front of you. Step your feet slightly wider than hip-width apart. Keeping your right leg and pelvis fixed, slowly internally rotate your left thigh so that your left knee moves toward your right leg (figure 9.48). Spread your left toes so that your left foot and ankle remain active. Slowly move your left leg back to a neutral position and repeat this movement with your right leg. Repeat for five cycles and then rest.

RESISTED HIP FLEXION

FIGURE 9.49   Resisted hip flexion.

Sit back in your chair and step your feet roughly hip-width apart again. Keeping your right knee bent, lift your right leg up a couple of inches and then gently press down into your right thigh with your hands to resist this movement (figure 9.49). Keep pressing your right thigh up into your hands. Hold this position for a few moments and then slowly lower your right foot back to the floor. Repeat with your left leg.

HAMSTRING STRENGTHENING

FIGURE 9.50   Hamstring strengthening.

Turn your whole body to the right so that you are sitting sideways on your chair with your left side in front. Aim to draw your left heel toward your left buttock, either using your left hand or your yoga strap or blanket. Release the support of your foot but try to hold your foot in this position for a few moments (figure 9.50). Experiencing a cramp in your hamstrings along the back of your thigh is very normal here. Remain connected to your breath and slowly lower your left foot to the floor. Take a moment to rest and then turn in your chair to face the opposite direction so that you can repeat this exercise with your right leg.

HAMSTRING STRETCHING

FIGURE 9.51   Hamstring stretching.

Turn back so that you are sitting in the chair facing forward again. Straighten your left leg and lift it up off the floor either using the support of your hands clasped under your left thigh or by looping your yoga strap around the sole of your left foot. Reach your left heel forward and gently draw the top of your left foot back toward you (figure 9.51). After a few moments, slowly lower your left foot, then repeat with your right leg.

SEATED PIGEON POSE (KAPOTASANA)

FIGURE 9.52   Seated Pigeon Pose.

Bend your right knee in toward your chest and then place the outer edge of your right foot onto your left thigh (figure 9.52). You have the option to step your left foot slightly forward in order to lower the height of your left thigh. Gently spread your right toes and press the big toe edge of your right foot farther to the left. You also have the option to hinge forward at your hip joints and lean your torso toward your legs. Take a couple of deep breaths and slowly raise your torso back up if you chose to lean forward. Challenge yourself to lift your right foot up away from your left thigh for just a moment and then release your right foot to the floor. Repeat this with your left leg, noticing how the left feels compared to the right.

ANKLE MOBILITY

FIGURE 9.53   Ankle mobility.

Hug your left knee against your chest and support your left leg with your hands. Slowly make circles with your left ankle and foot (figure 9.53), imagining that you are drawing a large oval shape with your big toe. Change the direction of the movements after a few repetitions and then repeat these movements with your right knee, foot, and ankle.

SUN SALUTATION (SURYA NAMASKAR)

Bring your hands into prayer position in front of your chest. As you inhale, reach your arms up and look up (figure 9.54a). As you exhale, gently fold forward over your legs and reach your hands toward the floor (figure 9.54b). With your next inhalation, lengthen your spine forward and look forward (figure 9.54c). As you exhale, bring your torso vertical and actively reach your arms forward (figure 9.54d), imagining that you are pressing your palms into a wall in front of you. Take a full inhalation, and as you exhale, slowly bend your elbows, drawing them back along the sides of your waist (figure 9.54e). With your next inhalation, press your hands down into your thighs and lift through the back of your chest, looking toward the ceiling (figure 9.54f). As you exhale, straighten your legs, lifting your feet up off the floor, and slowly raise your arms up in front of you (figure 9.54g). Take a couple of breaths here. On an inhalation, lower your feet back under your knees, fold forward over your legs, and lengthen your spine forward (figure 9.54h). As you then exhale, fold farther over your legs and release your hands toward the floor (figure 9.54i). As you then inhale, press down into your feet as you raise your torso back up and raise your arms above your head (figure 9.54j). With your exhalation, lower your hands back to prayer position in front of your chest (figure 9.54k).

FIGURE 9.54 Seated Sun Salutation: (a) arms reach up; (b) Seated Forward Fold; (c) spine lengthens forward; (d) Seated Plank;

(continued)

FIGURE 9.54 *(continued)   (e)* Seated Low Plank; *(f)* Seated Upward Facing Dog; *(g)* Seated Downward Facing Dog; *(h)* spine extends forward; *(i)* Seated Forward Fold; *(j)* arms reach up; *(k)* prayer position.

WARRIOR 2 (VIRABHADRASANA 2)

FIGURE 9.55   Warrior 2.

Turn your whole body to the right so that you come to sit sideways on your chair with your left side in front. Keep your right knee bent and extend your left leg back behind you, adjusting your position so that you can place your left foot securely on the floor. You have the option here to lift your arms out to the sides at roughly shoulder height while gazing at your right hand (figure 9.55) or place your hands on your hips. After a few breaths, slowly pivot on your heels so that you turn to face the left to repeat the pose on the other side, bending your left leg and straightening through your right leg.

TREE POSE (VRIKSHASANA)

FIGURE 9.56   Tree Pose.

Come back to your normal seated position and move slightly closer to the front of your chair. Straighten through your left leg and place your right foot anywhere along the length of your left leg. Gently press your left leg into your right foot and your right foot into your left leg (figure 9.56). You have the option to place your hands on your hips, in prayer position in front of your chest, or to reach your arms up toward the ceiling. You may also want to challenge yourself further by either looking up or closing your eyes. After a few breaths, slowly return to your neutral seated position and repeat these steps with your left side.

EAGLE POSE (GARUDASANA)

FIGURE 9.57   Eagle Pose.

Reach both of your arms out to the side and then cross them in front of your chest with your left arm on top of your right arm. You have the option to place your hands on your opposite shoulders, to turn your palms to face you, to interlace your thumbs or to place your palms together in prayer position. Lift your elbows as high as is comfortable and imagine that you are trying to gently pull your arms away from each other. Shift your weight into your left hip and cross your right leg over the top of the left (figure 9.57). You have the option here to try to gently press your outer right foot against your left leg or to hook your right foot behind your left calf. Keep your gaze forward, look up, or close your eyes. After a few breaths, unhook your arms and your legs and repeat these steps on the opposite side.

CHILD'S POSE (BALASANA)

FIGURE 9.58   Child's Pose.

Place a bolster or a stack of cushions on top of your thighs. Create enough height so that you can then comfortably lean forward and support the weight of your arms and your head onto the props (figure 9.58). If it feels good, you can gently roll your head side to side, and then rest for a few minutes, observing the gentle expansion and contraction of the back of your rib cage as you inhale and exhale. Slowly support your torso and come back up to a vertical position.

SEATED TWIST (MATSYENDRASANA)

FIGURE 9.59   Seated Twist.

Slowly twist your torso to the left, leading with your breastbone (figure 9.59). Your right knee can move an inch or two farther forward of your left knee. If it feels comfortable, slowly look over your left shoulder. You have the option to use the support of the chair back to help you to stay in the position for a few moments or rest your arms by your side. Slowly twist back to the center and then twist your torso in the opposite direction.

FINAL RELAXATION (SAVASANA)

FIGURE 9.60   Savasana.

Gently wrap a thin blanket over the back of your neck, cross it over your chest and tuck your elbows into the sides of your waist to secure the ends in place (figure 9.60). Once you are comfortable, you can either close your eyes or simply soften your gaze and focus your attention on something on the floor in front of you. Release your jaw and let your lips and teeth gently part. Allow your tongue to drop away from the roof of your mouth. Let go of any effort from your breath. Take a moment to scan your body with your mind, observing how you are feeling after the practice. Each time your mind wanders off, come back to focus on the gentle rise and fall of your inhalations and exhalations. After a few minutes, slowly begin to deepen your breath and bring gentle movement back into your fingers and your toes. Bring your hands into prayer position at the front of your chest. Use this time as an opportunity to set an intention for the rest of your day. Take a deep, full inhalation and as you exhale, bow your head toward your hands in honor of the yoga practice. If you closed your eyes, gently blink them back open.

RESTORATIVE YOGA PRACTICE

Restorative yoga is essentially a floor-based practice where asanas are held for longer than in conventional practices, often with the full support of props such as folded blankets and cushions or bolsters. The main aim of a restorative yoga practice is to relax the body and the mind, allowing the parasympathetic nervous system to dominate and therefore providing us with all the associated benefits that we have explored. To achieve this, it is important to make any necessary adjustments to feel comfortable in each position.

For this particular sequence, you will need a yoga mat, a sturdy chair, a bolster, a blanket, a towel, two foam bricks, an eye pillow, and a yoga strap. Again, you can get creative with props and use household items as substitutions.

We start the practice with a short meditation and breath work, followed by some gentle movement, then a short series of longer held asanas.

EFFORTLESS REST POSE

FIGURE 9.61 Effortless Rest Pose.

We begin this practice in a reclined position. Lie flat on your back if this feels comfortable for you. Bend your knees and place your feet under your knees, either keeping your feet roughly hip-width apart or stepping your feet wider and letting your inner knees rest against each other (figure 9.61). You can place a folded blanket or thin cushion underneath your head and neck. If lying flat does not feel comfortable, you can prop yourself up with your bolster and cushions. Rest your arms wherever they feel most comfortable. Once you have found your comfortable position, you can either close your eyes or simply soften your gaze and gently focus on a fixed point on the ceiling. Unhinge your jaw and let your lips and teeth gently part. Allow your tongue to drop away from the roof of your mouth. Take a moment to scan your body with your mind, observing how you are feeling today, physically, mentally, and emotionally. Try not to judge the sensations that you notice but simply observe them. Observe your breath for a few moments, noticing its quality. Take a moment to observe your mind, noticing if there are any particular thoughts at the forefront of your mind today.

BREATH WORK

FIGURE 9.62 Breath Work.

Place your hands on your lower abdomen so that your little fingers rest in your hip creases (figure 9.62). Begin to notice your abdomen gently expand as you inhale and gently fall as you exhale. If you notice your mind has wandered off, gently draw your focus back to the rise and fall of your abdomen. Begin now to focus your attention on the exhalations. Without straining, allow each exhalation to gently lengthen and then pause for a few moments at the end of the exhalation. Allow each inhalation to arise naturally without any effort and then focus once again on the exhalation. Repeat this for a few minutes.

We now do a visualization practice of alternative nostril breathing (nadi shodhana pranayama).

Exhale completely and then visualize inhaling through your left nostril only. Now visualize exhaling through your right nostril only. Visualize inhaling through your right nostril only and then exhaling through your left nostril. This presents one cycle of alternative nostril breathing.

Continue this practice for up to five minutes and complete the practice by finishing with an exhalation through your left nostril.

WINDSHIELD WIPERS POSE (SUPTA MATSYENDRASANA VARIATION)

FIGURE 9.63 Windshield Wipers Pose.

From Effortless Rest position, step your feet wider than hip-width apart, near the edges of your mat. You can reach your arms wide out to the sides or keep your hands resting on your abdomen. Take a full inhalation and as you exhale, slowly allow your legs to tip over to the left, spreading your toes to keep your feet active (figure 9.63). On your next inhalation, press the edges of your feet down and slowly draw your legs back to center. During your next exhalation, slowly allow your legs to tip over to the right. As you inhale, actively draw your legs back to the center. Repeat this sequence for a couple more rounds with the option of turning your head to look in the opposite direction that your legs are moving.

RECLINING HAND-TO-BIG-TOE POSE (SUPTA PADANGUSTHASANA)

a

b

FIGURE 9.64 Reclining Hand-to-Big-Toe *(a)* Pose A; *(b)* Pose B.

Staying in Effortless Rest position, place your feet roughly hip-width apart. Hug your right knee against your chest and loop the center of your yoga strap around the arch of your right foot. Hold one end of the strap in each hand and slowly move your foot away from your chest until you can straighten your leg (maintain a slight bend in your knee if you prefer). Let your hand slide down the strap until you can comfortably rest your elbows on the floor. Gently press the heel of your right foot away from you as you draw the top of your right foot back toward you (figure 9.64a). You have the option here to keep your left foot flat on the floor or to straighten your left leg out on the mat, pressing it down into the mat to keep you anchored. Hold this position for five breath cycles.

Now take hold of both ends of your yoga strap with your right hand and place your left hand on your left hip. Keeping the left side of your pelvis grounded, slowly externally rotate your right hip, turning your toes out slightly to the right. With control, slowly lower your right leg to the right while keeping the left side of your pelvis grounded (figure 9.64b). At the limit of your movement, pause for a couple of breaths and then actively draw your right leg back to the center and slowly lower your right leg to the mat to rest. Stretch your legs on the mat for a moment and notice how your legs are feeling. When you are ready, repeat this sequence with the opposite side, noticing how your left leg and hip feel compared to your right side.

LEGS-UP-CHAIR POSE (VIPARITA KARANI VARIATION)

FIGURE 9.65 Legs-Up-Chair Pose.

Lie on your side on the mat, close to the edge of your chair. Slowly roll onto your back, hugging your knees close to your chest and then place your legs onto the chair (figure 9.65). You have the option here to use your sofa instead of a chair if that works better for you. You can also choose to elevate your pelvis by placing your bolster or folded blanket under it. Rest your arms by your side or place your hands on your abdomen. Have a blanket close by in case you feel cold, and feel free to use an eye pillow with this pose, too. Remain in this position for up to five minutes and then slowly move your feet back to the floor.

INCLINE TWIST (SUPTA MATSYENDRASANA VARIATION)

FIGURE 9.66   Incline Twist.

For this asana you need to create an incline with your props. You can experiment by placing your foam bricks on different edges to give you different heights. Place your bolster along the incline and sit on your right side with your right hip bone close to the lower end of your bolster. Keep your legs bent and play with the positioning of your legs until you find a comfortable option. You can place a blanket between your legs for more support. Slowly twist your torso to the right so that your chest faces the bolster, and then lean down onto the incline that you have created (figure 9.66). You can choose to keep your head facing the direction that your knees are pointing or turn your head to look in the opposite direction. Once you are comfortable, either close your eyes or simply soften your gaze. Remain in this position for up to five minutes and then very slowly twist back to your original position, ready to move to the opposite side.

RECLINING BOUND ANGLE POSE (SUPTA BADDHA KONASANA)

FIGURE 9.67   Reclining Bound Angle Pose.

Keeping the incline that you have created with your props, sit with your back to the incline just a couple of inches away from the lower end of the bolster. With your knees bent and feet flat on the floor, slowly lower yourself so that you are completely supported by the props. If the incline feels too flat or too steep, adjust. Step your feet closer together and then spread your knees apart (figure 9.67). Consider rolling up a long blanket, looping it over the top of your feet, under your outer thighs, and tucking it into the side of your waist. Use your eye pillow here if that feels comfortable and lean into this position for up to 10 minutes. Slowly draw your knees together and take your time to roll over onto your side. Move the props to the side and roll onto your back for a few moments. Notice how you feel.

PRONE RELAXATION (SAVASANA)

FIGURE 9.68   Prone Relaxation (Savasana).

We finish with Savasana lying on our front (figure 9.68). This does not feel comfortable for everyone, so you also have the option to take any position that feels most restful. Lay your bolster lengthwise along the center of your mat and place a foam brick in front of the top end of your bolster with a thin, folded blanket or towel on top of it. At the bottom section of your mat, lay another blanket and roll up the bottom end slightly. Come to lie facedown with your hip creases at the lower edge of your bolster. Adjust the foam brick so that you can comfortably rest your forehead down with an open space for the rest of your face. Support the front of your ankles with the rolled-up end of your blanket. Remain in this position for up to 10 minutes.

To close, slowly ease yourself up to a comfortable sitting position. Take a moment to reflect on how you feel after your practice. Observe your breath for a few moments. Then, bring your hands into prayer position at the front of your chest. Use this time as an opportunity to set an intention for the rest of your day. Take a deep, full inhalation and as you exhale, bow your head toward your hands in honor of the yoga practice. If you closed your eyes, gently blink them back open.

CONCLUSION

Raja yoga, which includes asana and pranayama, is but one path to liberation according to the Upanishads and Bhagavad Gita. These texts also describe jñāna yoga, or the pursuit of knowledge, as another path. The central mission of jñāna yoga is to know oneself, to understand one's true nature.

Learning physiology of the human body is one part of learning about one's true self. By understanding our bodies, which we are so very blessed to inhabit in this life, we learn how to better care for them. Though some of the historical claims about yoga's physiological benefits may be rather far-flung, yoga's many benefits are nonetheless very real and each practitioner probably has their own story of how yoga has positively affected their life. We hope this book has helped to deepen your own practice through self-study, or jñāna yoga, and that you feel inspired to share the benefits of yoga with others, as yoga is truly for all.

References

Introduction

Paul, R., and L. Elder. 2019. *The Miniature Guide to Critical Thinking Concepts & Tools.* Washington, DC: Rowman & Littlefield.

Russell, B. 1929. *Marriage and Morals.* London: George Allen and Unwin; New York: Horace Liveright.

Sackett, D., S. Straus, W. Richardson, W. Rosenberg, and R. Haynes. 2000. *Evidence-Based Medicine.* Philadelphia: Churchill Livingstone.

Chapter 1

Adstrum, S., G. Hedley, R. Schleip, C. Stecco, and C.A. Yucesoy. 2017. "Defining the Fascial System." *Journal of Bodywork and Movement Therapies* 21 (1): 173-177.

Airaksinen, O., J. Brox, C. Cedraschi, J. Hildebrandt, J. Klaber-Moffett, F. Kovacs, et al. 2006. "Chapter 4. European Guidelines for the Management of Chronic Nonspecific Low Back Pain." *European Spine Journal* 15 (Suppl. 2): S192-S300.

Ajimsha, M.S., N.R. Al-Mudahka, and J.A. Al-Madzhar. 2015. "Effectiveness of Myofascial Release: Systematic Review of Randomized Controlled Trials." *Journal of Bodywork and Movement Therapies* 19 (1): 102-112. https://doi.org/10.1016/j.jbmt.2014.06.001.

Baars, J.H., R. Mager, K. Dankert, M. Hackbarth, F. von Dincklage, and B. Rehberg. 2009. "Effects of Sevoflurane and Propofol on the Nociceptive Withdrawal Reflex and on the H Reflex." *Anesthesiology* 111:72-81. https://doi.org/10.1097/ALN.0b013e3181a4c706.

Beales, D., A. Smith, and P. O'Sullivan, et al. 2015. "Back Pain Beliefs Are Related to the Impact of Low Back Pain in Baby Boomers in the Busselton Healthy Aging Study." *Physical Therapy* 95:180-189. https://doi.org/10.2522/ptj.20140064.

Beardsley, C. 2020. "What Causes Delayed Onset Muscle Soreness (DOMS)?" *Medium,* June 8, 2020. https://medium.com/@SandCResearch/what-causes-delayed-onset-muscle-soreness-doms-d126d04bbb3a.

Behm, D.G., A.J. Blazevich, A.D. Kay, and M. McHugh. 2016. "Acute Effects of Muscle Stretching on Physical Performance, Range of Motion, and Injury Incidence in Healthy Active Individuals: A Systematic Review." *Applied Physiology, Nutrition, and Metabolism* 41 (1): 1-11.

Behm, D.G., and J. Wilke. 2019. "Do Self-Myofascial Release Devices Release Myofascia? Rolling Mechanisms: A Narrative Review." *Sports Medicine* 49:1173-1181. https://doi.org/10.1007/s40279-019-01149-y.

Benedetti, F., M. Lanotte, L. Lopiano, and L. Colloca. 2007. "When Words Are Painful: Unraveling the Mechanisms of the Nocebo Effect." *Neuroscience* 147 (2): 260-271. https://doi.org/10.1016/j.neuroscience.2007.02.020.

Biswas, A., P.I. Oh, G.E. Faulkner, et al. 2015. "Sedentary Time and Its Association With Risk for Disease Incidence, Mortality, and Hospitalization in Adults: A Systematic Review and Meta-Analysis" [published correction appears in *Annals of Internal Medicine* 163 (5): 400]. *Annals of Internal Medicine* 162 (2): 123-132. https://doi.org/10.7326/M14-1651.

Chalmers, G. 2004. "Strength Training: Re-examination of the Possible Role of Golgi Tendon Organ and Muscle Spindle Reflexes in Proprioceptive Neuromuscular Facilitation Muscle Stretching." *Sports Biomechanics* 3 (1): 159-183.

Chan, S.C.W., S.J. Ferguson, and B. Gantenbein-Ritter. 2011. "The Effects of Dynamic Loading on the Intervertebral Disc." *European Spine Journal* 20 (11): 1796-1812.

Clark, B. 2012. *The Complete Guide to Yin Yoga: The Philosophy and Practice of Yin Yoga.* Vancouver: Wild Strawberry Productions.

Clinch J., K. Deere, A. Sayers, S. Palmer, C. Riddoch, J.H. Tobias, and E.M. Clark. 2011. "Epidemiology of Generalized Joint Laxity (Hypermobility) in Fourteen-Year-Old Children From the UK: A Population-Based Evaluation." *Arthritis & Rheumatism* 63(9): 2819-2827.

Cooper, C., G. Campion, and L.J. Melton, 3rd. 1992. "Hip Fractures in the Elderly: A World-Wide Projection." *Osteoporosis International: A Journal Established as Result of Cooperation Between the European Foundation for Osteoporosis and the National Osteoporosis Foundation of the USA* 2 (6): 285-289. https://doi.org/10.1007/BF01623184.

Cramer, H., C. Krucoff, and G. Dobos. 2013. "Adverse Events Associated With Yoga: A Systematic Review of Published Case Reports and Case Series." *PLOS ONE* 8 (10): e75515.

Cramer, H., T. Ostermann, and G. Dobos. 2018. "Injuries and Other Adverse Events Associated With Yoga Practice: A Systematic Review of Epidemiological Studies." *Journal of Science and Medicine in Sport* 21 (2): 147-154.

Cramer, H., L. Ward, R. Saper, D. Fishbein, G. Dobos, and R. Lauche. 2015. "The Safety of Yoga: A Systematic Review and Meta-Analysis of Randomized Controlled Trials." *American Journal of Epidemiology* 182 (4): 281-293. https://doi.org/10.1093/aje/kwv071.

Data-Franco, J., and M. Berk. 2013. "The Nocebo Effect: A Clinician's Guide." *Australian and New Zealand Journal of Psychiatry* 47:617-623.

Davis, D.S., P.E. Ashby, K.L. McCale, J.A. McQuain, and J.M. Wine. 2005. "The Effectiveness of 3 Stretching Techniques on Hamstring Flexibility Using Consistent Stretching Parameters." *The Journal of Strength and Conditioning Research* 19 (1): 27-32. https://doi.org/10.1519/14273.1.

Diamond, T.H., S.W. Thornley, R. Sekel, and P. Smerdely. 1997. "Hip Fracture in Elderly Men: Prognostic Factors and Outcomes." *The Medical Journal of Australia* 167 (8): 412-415. https://doi.org/10.5694/j.1326-5377.1997.tb126646.x.

DiGiovanna, E.L., S. Schiowitz, and D.J. Dowling, eds. 2005. *An Osteopathic Approach to Diagnosis and Treatment.* Philadelphia: Lippincott Williams & Wilkins.

Drici, M-D., F. Raybaud, C. Lunardo, P. Iacono, and P. Gustovic. 1995. "Influence of the Behaviour Pattern on the Nocebo Response of Healthy Volunteers." *British Journal of Clinical Pharmacology* 39:204-206.

Dupuy, O., W. Douzi, D. Theurot, L. Bosquet, and B. Dugué. 2018. "An Evidence-Based Approach for Choosing Post-Exercise Recovery Techniques to Reduce Markers of Muscle Damage, Soreness, Fatigue, and Inflammation: A Systematic Review With Meta-Analysis." *Frontiers in Physiology* 9 (April 26): 403. https://doi.org/10.3389/fphys.2018.00403.

Freitas, S.R., B. Mendes, G. Le Sant, et al. 2018. "Can Chronic Stretching Change the Muscle-Tendon Mechanical Properties? A Review." *Scandinavian Journal of Medicine & Science in Sports* 28 (3): 794-806.

Frost, H.M. 1964. *The Laws of Bone Structure.* Springfield, IL: Thomas.

Garcia-Campayo, J., E. Asso, and M. Alda. 2011. "Joint Hypermobility and Anxiety: The State of the Art." *Current Psychiatry Reports* 13 (1): 18-25. https://doi.org/10.1007/s11920-010-0164-0.

Genant, H.K., C. Cooper, G. Poor, I. Reid, G. Ehrlich, J. Kanis, B.E.C. Nordin, et al. 1999. "Interim Report and Recommendations of the World Health Organization Task-Force for Osteoporosis." *Osteoporosis International* 10 (4): 259.

Giesser, B.S. 2015. "Exercise in the Management of Persons With Multiple Sclerosis." *Therapeutic Advances in Neurological Disorders* 8 (3): 123-130. https://doi.org/10.1177/175628561576663.

Gmada, N., E. Bouhlel, I. Mrizak, H. Debabi, M. Ben Jaballah, Z. Tabka, Y. Feki, and M. Amri. 2005. "Effect of Combined Active Recovery From Supramaximal Exercise on Blood Lactate Disappearance in Trained and Untrained Man." *International Journal of Sports Medicine* 26 (10): 874-879.

Pratelli, E., I. Cinotti, and P. Pasquetti. 2010. "Rehabilitation in Osteoporotic Vertebral Fractures." *Clinical Cases in Mineral and Bone Metabolism* 7 (1): 45.

Quintner, J.L., and Cohen, M.L. 1994. "Referred Pain of Peripheral Nerve Origin: An Alternative to the 'Myofascial Pain' Construct." *Clinical Journal of Pain* 10 (3): 243-251. https://doi.org/10.1097/00002508-199409000-00012.

Reeves, N.D. 2006. "Adaptation of the Tendon to Mechanical Usage." *Journal of Musculoskeletal & Neuronal Interactions* 6 (2): 174-180.

Robling, A.G., A.B. Castillo, and C.H. Turner. 2006. "Biomechanical and Molecular Regulation of Bone Remodeling. *Annual Review of Biomedical Engineering* 8:455-498. https://doi.org/10.1146/annurev.bioeng.8.061505.095721.

Russo, C.R. 2009. "The Effects of Exercise on Bone. Basic Concepts and Implications for the Prevention of Fractures." *Clinical Cases in Mineral Bone Metabolism* 6 (3): 223-228.

Ryan, E.D., T.J. Herda, P.B. Costa, A.A. Walter, K.M. Hoge, J.R. Stout, and J.T. Cramer. 2010. "Viscoelastic Creep in the Human Skeletal Muscle-Tendon Unit." *European Journal of Applied Physiology* 108 (1): 207-211.

Scheper, M.C., J.E. de Vries, J. Verbunt, and R.H. Engelbert. 2015. "Chronic Pain in Hypermobility Syndrome and Ehlers-Danlos Syndrome (Hypermobility Type): It Is a Challenge." *Journal of Pain Research* 20 (8): 591-601. https://doi.org/10.2147/JPR.S64251.

Schleip, R., W. Klingler, and F. Lehmann-Horn. 2005. "Active Fascial Contractility: Fascia May Be Able to Contract in a Smooth Muscle-Like Manner and Thereby Influence Musculoskeletal Dynamics." *Medical Hypotheses* 65:273-277.

Sharman, M.J., A.G. Cresswell, and S. Riek. 2006. "Proprioceptive Neuromuscular Facilitation Stretching: Mechanisms and Clinical Implications. *Sports Medicine* 36 (11): 929-939. https://doi.org/10.2165/00007256-200636110-00002.

Shields, B.J., and G.A. Smith. 2009. "Cheerleading-Related Injuries in the United States: A Prospective Surveillance Study." *Journal of Athletic Training* 44 (6): 567-577.

Sinaki, M. 2007. "The Role of Physical Activity in Bone Health: A New Hypothesis to Reduce Risk of Vertebral Fracture." *Physical Medicine and Rehabilitation Clinics of North America* 18 (3): 593-608.

Sinaki, M. 2013. "Yoga Spinal Flexion Positions and Vertebral Compression Fracture in Osteopenia or Osteoporosis of Spine: Case Series." *Pain Practice* 13 (1): 68-75.

Sinaki, M., E. Itoi, H.W. Wahner, P. Wollan, R. Gelzcer, B.P. Mullan, D.A. Collins, and S.F. Hodgson. 2002. "Stronger Back Muscles Reduce the Incidence of Vertebral Fractures: A Prospective 10 Year Follow-Up of Postmenopausal Women." *Bone* 30 (6): 836-841.

Sinaki, M., and B.A. Mikkelsen. 1984. "Postmenopausal Spinal Osteoporosis: Flexion Versus Extension Exercises." *Archives of Physical Medicine and Rehabilitation* 65 (10): 593-596.

Smith, E.N., and A. Boser. 2013. "Yoga, Vertebral Fractures, and Osteoporosis: Research and Recommendations." *International Journal of Yoga Therapy* 23 (1): 17-23.

Souza, T.R., S.T. Fonseca, G.G. Gonçalves, J.M. Ocarino, and M.C. Mancini. 2009. "Prestress Revealed by Passive Co-Tension at the Ankle Joint." *Journal of Biomechanics* 42 (14): 2374-2380. https://doi.org/10.1016/j.jbiomech.2009.06.033.

Standring, S., ed. 2004. *Gray's Anatomy*. 39th ed. London: Churchill Livingstone.

Stanton, T.R., G.L. Moseley, A.Y.L. Wong, and G.N. Kawchuk. 2017. "Feeling Stiffness in the Back: A Protective Perceptual Inference in Chronic Back Pain." *Scientific Reports* 7 (1): 9681. https://doi.org/10.1038/s41598-017-09429-1.

Taleb, N.N. 2012. *Antifragile: Things That Gain From Disorder*. Vol. 3. New York: Random House Incorporated.

Tinkle, B., M. Castori, B. Berglund, H. Cohen, R. Grahame, H. Kazkaz, and H. Levy. 2017. "Hypermobile Ehlers-Danlos Syndrome (a.k.a. Ehlers-Danlos Syndrome Type III and Ehlers-Danlos Syndrome Hypermobility Type): Clinical Description and Natural History." *American Journal of Medical Genetics C Seminars in Medical Genetics* 175 (1): 48-69. https://doi.org/10.1002/ajmg.c.31538.

Travell, J.G., and Simons, D.G. 1983. *Myofascial Pain and Dysfunction: The Trigger Point Manual.* Baltimore: Williams & Willkins.

Turner, C.H., and F.M. Pavalko. 1998. "Mechanotransduction and Functional Response of the Skeleton to Physical Stress: The Mechanisms and Mechanics of Bone Adaptation." *Journal of Orthopaedic Science: Official Journal of the Japanese Orthopaedic Association* 3 (6): 346-355. https://doi.org/10.1007/s007760050064.

U.S. Department of Health and Human Services. 2004. *Bone Health and Osteoporosis: A Report of the Surgeon General.* Rockville, MD: U.S. Department of Health and Human Services, Office of the Surgeon General.

Våben, C., K.M. Heinemeier, P. Schjerling, J. Olsen, M.M. Petersen, M. Kjaer, and M.R. Krogsgaard. 2020. "No Detectable Remodelling in Adult Human Menisci: An Analysis Based on the C14 Bomb Pulse." *British Journal of Sports Medicine* 54 (23): 1433-1437. https://doi.org/10.1136/bjsports-2019-101360.

Vogt, M., and H.H. Hoppeler. 2014. "Eccentric Exercise: Mechanisms and Effects When Used as Training Regime or Training Adjunct." *Journal of Applied Physiology (1985)* 116(11): 1446-1454.

Waddell, G. 1987. "1987 Volvo Award in Clinical Sciences. A New Clinical Model for the Treatment of Low-Back Pain." *Spine* 12 (7): 632-644.

Wells, R.E., and T.J. Kaptchuk. 2012. "To Tell the Truth, the Whole Truth, May Do Patients Harm: The Problem of the Nocebo Effect for Informed Consent." *American Journal of Bioethics* 12:22-29.

Weppler, C.H., and S.P. Magnusson. 2010. "Increasing Muscle Extensibility: A Matter of Increasing Length or Modifying Sensation?" *Physical Therapy* 90 (3): 438-449. https://doi.org/10.2522/ptj.20090012.

Wiese, C., D. Keil, A.S. Rasmussen, and R. Olesen. 2019. "Injury in Yoga Asana Practice: Assessment of the Risks." *Journal of Bodywork and Movement Therapies* 23 (3): 479-488.

Wiewelhove, T., A. Döweling, C. Schneider, L. Hottenrott, T. Meyer, M. Kellmann, M. Pfeiffer, and A. Ferrauti. 2019. "A Meta-Analysis of the Effects of Foam Rolling on Performance and Recovery." *Frontiers in Physiology* 9 (10): 376. https://doi.org/10.3389/fphys.2019.00376.

Wolfe, F. 2013. "Travell, Simons and Cargo Cult Science." *The Fibromyalgia Perplex* (blog), February 19, 2013. www.fmperplex.com/2013/02.

Wolfe, F., D.G., Simons, J. Fricton, R.M. Bennett, D.L. Goldenberg, R. Gerwin, D. Hathaway, G.A. McCain, I.J. Russell, H.O. Sanders, et al. 1992. "The Fibromyalgia and Myofascial Pain Syndromes: A Preliminary Study of Tender Points and Trigger Points in Persons With Fibromyalgia, Myofascial Pain Syndrome and No Disease." *Journal of Rheumatology* 19 (6): 944-951.

Wolff, J., trans. 1986. *The Law of Bone Remodeling* (translated from the 1892 original, *Das Gesetz der Transformation der Knochen*, by P. Maquet and R. Furlong). Berlin: Springer Verlag.

Yoga Alliance. 2016. "Highlights From the 2016 Yoga in America Study." www.yogaalliance.org/Learn/About_Yoga/2016_Yoga_in_America_Study/Highlights.

Chapter 2

Almeida, D., S. Charles, J. Mogle, J. Drewelies, C. Aldwin, A. Spiro III, and D. Gerstorf. 2020. "Charting Adult Development Through (Historically Changing) Daily Stress Processes." *American Psychologist* 75 (4): 511-524.

American Psychiatric Association. 2013. *The Diagnostic and Statistical Manual of Mental Disorders, Fifth Edition (DSM–5).* Washington, DC: American Psychiatric Publishing.

Applegate, C., B. Kapp, M. Underwood, and C. McNall. 1983. "Autonomic and Somatomotor Effects of Amygdala Central N. Stimulation in Awake Rabbits." *Physiology and Behavior* 31 (3): 353-630.

Barlow, D. 2002. *Anxiety and Its Disorders: The Nature and Treatment of Anxiety and Panic.* New York: Guilford Press.

Berger, M., J. Gray, and B. Roth. 2009. "The Expanded Biology of Serotonin." *Annual Review of Medicine* 60:355-366.

Bernardi, L., P. Sleight, G. Bandinelli, S. Cencetti, L. Fattorini, J. Wdowczyc-Szulc, and A. Lagi. 2001. "Effect of Rosary Prayer and Yoga Mantras on Autonomic Cardiovascular Rhythms: Comparative Study." *British Medical Journal (Clinical Research Edition)* 323 (7327): 1446-1449.

Biggs, E., A. Meulders, and J. Vlaeyen. 2016. "The Neuroscience of Pain and Fear." In *Neuroscience of Pain, Stress, and Emotion*, edited by M. al'Absi and M. Arve Flaten.148-162. Cambridge, MA: Academic Press.

Borrell-Carrió, F., A. Suchman, and R. Epstein. 2004. "The Biopsychosocial Model 25 Years Later: Principles, Practice, and Scientific Inquiry." *Annals of Family Medicine* 2 (6): 576-582.

Bougea, A. 2020. "An Evaluation of the Studies on the Therapeutic Effects of Yoga in People With Dementia." *EMJ Neurology* 8 (1): 64-66.

Brenes, G., S. Sohl, R. Wells, D. Befus, C. Campos, and S. Danhauer. 2019. "The Effects of Yoga on Patients With Mild Cognitive Impairment and Dementia: A Scoping Review." *American Journal of Geriatric Psychiatry* 27 (2): 188-197.

Brinsley, J., F. Schuch, O. Lederman, D. Girard, M. Smout, M. Immink, B. Stubb, J. Firth, K. Davison, and S. Rosenbaum. 2020. "Effects of Yoga on Depressive Symptoms in People With Mental Disorders: A Systematic Review and Meta-Analysis." *British Journal of Sports Medicine* 55(17): 992-1000.

Cadegiani, F., and C. Kater. 2016. "Adrenal Fatigue Does Not Exist: A Systematic Review." *BMC Endocrine Disorders* 16 (1): 48.

Cannon, W. 1915. *Bodily Changes in Pain, Hunger, Fear and Rage: An Account of Recent Researches Into the Function of Emotional Excitement.* New York: Appleton and Company.

Carabotti, M., A. Scirocco, M. Maselli, and C. Severi. 2015. "The Gut-Brain Axis: Interactions Between Enteric Microbiota, Central and Enteric Nervous Systems." *Annals of Gastroenterology* 28 (2): 203-209.

Charney, D., and H. Manji. 2004. "Life Stress, Genes, and Depression: Multiple Pathways Lead to Increased Risk and New Opportunities for Intervention." *Science STKE* 225 (re5).

Cherup, N., K. Strand, L. Lucchi, S. Wooten, C. Luca, and J. Signorile. 2020. "Yoga Meditation Enhances Proprioception and Balance in Individuals Diagnosed With Parkinson's Disease." *Perceptual and Motor Skills* 128(1): 304-323.

Chrousos, G. 2009. "Stress and Disorders of the Stress System." *Nature Reviews. Endocrinology* 5 (7): 374-381.

Cramer, H., D. Anheyer, F. Saha, and G. Dobos. 2018. "Yoga for Posttraumatic Stress Disorder—A Systematic Review and Meta-Analysis." *BMC Psychiatry* 18(1): 72.

Cramer, H., R. Lauche, H. Haller, and G. Dobos. 2013. "A Systematic Review and Meta-Analysis of Yoga for Low Back Pain." *The Clinical Journal of Pain* 29 (5): 450-460.

Cramer, H., R. Lauche, J. Langhorst, and G. Dobos. 2013. "Yoga for Depression: A Systematic Review and Meta-Analysis." *Depression Anxiety* 30:1068-1083.

Crofford, L. 2015. "Chronic Pain: Where the Body Meets the Brain." *Transactions of the American Clinical and Climatological Association* 126:167-183.

Daneman, R., and A. Prat. 2015. "The Blood–Brain Barrier." *Cold Spring Harbor Perspectives in Biology* 7(1): a020412.

Desai, R., A. Tailor, and T. Bhatt. 2015. "Effects of Yoga on Brain Waves and Structural Activation: A Review." *Complementary Therapies in Clinical Practice* 21 (2): 112-118.

De Zeeuw, C., and M. Ten Brinke. 2015. "Motor Learning and the Cerebellum." *Cold Spring Harbor Perspectives in Biology* 7(9): a021683.

Du, Q., and Z. Wei. 2017. "The Therapeutic Effects of Yoga in People With Dementia: A Systematic Review." *International Journal of Geriatric Psychiatry* 32 (1): 118.

Eagleman, D. 2020. *Livewired: The Inside Story of the Ever-Changing Brain*. New York: Pantheon.

Engel, G. 1977. "The Need for a New Medical Model: A Challenge for Biomedicine." *Science* 196 (4286): 129-136.

Fayad, F., M. Lefevre-Colau, S. Poiraudeau, and J. Fermanian. 2004. "Chronicity, Recurrence, and Return to Work in Low Back Pain: Common Prognostic Factors." *Annales de Réadaptation et de Médecine Physique* 47:179-189.

Gothe, N., I. Khan, J. Hayes, E. Erlenbach, and J. Damoiseaux. 2019. "Yoga Effects on Brain Health: A Systematic Review of the Current Literature." *Brain Plasticity* 5 (1): 105-122.

Gothe, N., and McAuley, E. 2015. "Yoga and Cognition: A Meta-Analysis of Chronic and Acute Effects." *Psychosomatic Medicine* 77(7): 784-797.

Gotink, R., M. Vernooij, M. Ikram, W. Niessen, G. Krestin, A. Hofman, H. Tiemeier, and M. Hunink. 2018. "Meditation and Yoga Practice Are Associated With Smaller Right Amygdala Volume: The Rotterdam Study." *Brain Imaging and Behavior* 12 (6): 1631-1639.

Greden, J. 2001. "The Burden of Recurrent Depression: Causes, Consequences, and Future Prospects." *Journal of Clinical Psychiatry* 62 (Suppl. 22): 5-9.

Green, E., A. Huynh, L. Broussard, B. Zunker, J. Matthews, C. Hilton, and K. Aranha. 2019. "Systematic Review of Yoga and Balance: Effect on Adults With Neuromuscular Impairment." *The American Journal of Occupational Therapy: Official Publication of the American Occupational Therapy Association* 73(1): 7301205150p1-7301205150p11.

He, W., X. Wang, H. Shi, H. Shang, L. Li, X. Jing, and B. Zhu. 2012. "Auricular Acupuncture and Vagal Regulation." *Evidence-Based Complementary and Alternative Medicine: eCAM* 2012: 786839.

Hilton, L., S. Hempel, B.A. Ewing, E. Apaydin, L. Xenakis, S. Newberry, B. Colaiaco, A. Ruelaz Maher, R.M. Shanman, M.E. Sorbero, and M.A. Maglione. 2017. "Mindfulness Meditation for Chronic Pain: Systematic Review and Meta-Analysis." *Annals of Behavioral Medicine* 51 (2): 199-213.

Hilton, L., A. Ruelaz Maher, B. Colaiaco, E. Apaydin, M. Sorbero, M. Booth, R. Shanman, and S. Hempel. 2017. "Meditation for Posttraumatic Stress: Systematic Review and Meta-Analysis." *Psychological Trauma: Theory, Research, Practice, and Policy* 9 (4): 453-460.

Holzel, B., J. Carmody, K. Evans, E. Hoge, J. Dusek, L. Morgan, R. Pitman, and S. Lazar. 2010. "Stress Reduction Correlates With Structural Changes in the Amygdala." *Social Cognitive and Affective Neuroscience* 5 (1): 11-17.

Interagency Pain Research Coordinating Committee. 2016. *National Pain Strategy: A Comprehensive Population Health-Level Strategy for Pain*. Washington, DC: U.S. Department of Health and Human Services, National Institutes of Health.

Iwata, J., K. Chida, and J. LeDoux. 1987. "Cardiovascular Responses Elicited by Stimulation of Neurons in the Central Amygdaloid Nucleus in Awake but Not Anesthetized Rats Resemble Conditioned Emotional Responses." *Brain Research* 418 (1): 183-188.

Janak, P., and K. Tye. 2015. "From Circuits to Behaviour in the Amygdala." *Nature* 517 (7534): 284-292.

Jans-Beken, L., N. Jacobs, M. Janssens, S. Peeters, J. Reijnders, L. Lechner, and J. Lataster. 2019. "Gratitude and Health: An Updated Review." *Journal of Positive Psychology* 15 (6): 743-782.

Jeter, P., A. Nkodo, S. Moonaz, and G. Dagnelie. 2014. "A Systematic Review of Yoga for Balance in a Healthy Population." *Journal of Alternative and Complementary Medicine* 20 (4): 221-232.

Jin, X., L. Wang, S. Liu, L. Zhu, P. Loprinzi, and X. Fan. 2019. "The Impact of Mind–Body Exercises on Motor Function, Depressive Symptoms, and Quality of Life in Parkinson's Disease: A Systematic Review and Meta-Analysis." *International Journal of Environmental Research and Public Health* 17 (1): 31.

Kalyani, B., G. Venkatasubramanian, R. Arasappa, N. Rao, S. Kalmady, R. Behere, H. Rao, M. Vasudev, and B. Gangadhar. 2011. "Neurohemodynamic Correlates of 'OM' Chanting: A Pilot Functional Magnetic Resonance Imaging Study." *International Journal of Yoga* 4 (1): 3-6.

Kim, S. 2016. "Effects of Yoga on Chronic Neck Pain: A Systematic Review of Randomized Controlled Trials." *Journal of Physical Therapy Science* 28 (7): 2171-2174.

Ko, J., and Strafella, A. 2012. "Dopaminergic Neurotransmission in the Human Brain: New Lessons From Perturbation and Imaging." *Neuroscientist* 18 (2): 149-168.

Kok, B., K. Coffey, M. Cohn, L. Catalino, T. Vacharkulksemsuk, S. Algoe, M. Brantley, and B. Fredrickson 2013. "How Positive Emotions Build Physical Health: Perceived Positive Social Connections Account for the Upward Spiral Between Positive Emotions and Vagal Tone." *Psychological Science* 24 (7): 1123-1132.

Lewis, T., F. Amini, and R. Lannon. 2007. *A General Theory of Love.* New York: Knopf Doubleday Publishing Group.

Ley, R., D. Peterson, and J. Gordon. 2006. "Ecological and Evolutionary Forces Shaping Microbial Diversity in the Human Intestine." *Cell* 124:837-848.

Li, A., and C. Goldsmith. 2012. "The Effects of Yoga on Anxiety and Stress." *Alternative Medicine Review* 17:21-35.

Lindahl, J., N. Fisher, D. Cooper, R. Rosen, and W. Britton. 2017. "The Varieties of Contemplative Experience: A Mixed-Methods Study of Meditation-Related Challenges in Western Buddhists." *PLOS ONE* 12 (5): e0176239.

Liu, D., C. Caldji, S. Sharma, P. Plotsky, and M. Meaney. 2000. "Influence of Neonatal Rearing Conditions on Stress-Induced Adrenocorticotropin Responses and Norepinepherine Release in the Hypothalamic Paraventricular Nucleus." *Journal of Neuroendocrinology* 12 (1): 5-12.

London, A., I. Benhar, and M. Schwartz. 2013. "The Retina as a Window to the Brain—From Eye Research to CNS Disorders." *Nature Reviews Neurology* 9:44-53.

Lu, W., G. Chen, and C. Kuo. 2011. "Foot Reflexology Can Increase Vagal Modulation, Decrease Sympathetic Modulation, and Lower Blood Pressure in Healthy Subjects and Patients With Coronary Artery Disease." *Alternative Therapies in Health and Medicine* 17 (4): 8-14.

Lurie, I., Y. Yang, K. Haynes, R. Mamtani, and B. Boursi. 2015. "Antibiotic Exposure and the Risk for Depression, Anxiety, or Psychosis: A Nested Case-Control Study." *Journal of Clinical Psychiatry* 76 (11): 1522-1528.

Mäkinen, T., M. Mäntysaari, T. Pääkkönen, J. Jokelainen, L. Palinkas, J. Hassi, J. Leppäluoto, K. Tahvanainen, and H. Rintamäki. 2008. "Autonomic Nervous Function During Whole-Body Cold Exposure Before and After Cold Acclimation." *Aviation, Space, and Environmental Medicine* 79 (9): 875-882.

Maniam, J., C. Antoniadis, and M. Morris. 2014. "Early-Life Stress, HPA Axis Adaptation, and Mechanisms Contributing to Later Health Outcomes." *Frontiers in Endocrinology* 5: 73.

McCorry, L. 2007. "Physiology of the Autonomic Nervous System." *American Journal of Pharmaceutical Education* 71 (4): 78.

Merskey, H., and N. Bogduk. 1994. *Classification of Chronic Pain: Descriptions of Chronic Pain Syndromes and Definitions of Pain Terms.* 2nd ed. Seattle: International Association for the Study of Pain (IASP) Press.

Moylan, S., M. Maes, N. Wray, and M. Berk. 2013. "The Neuroprogressive Nature of Major Depressive Disorder: Pathways to Disease Evolution and Resistance, and Therapeutic Implications." *Molecular Psychiatry* 18 (5): 595-606.

Neugebauer, V. 2015. "Amygdala Pain Mechanisms." *Handbook of Experimental Pharmacology* 227:261-284.

Neugebauer, V., W. Li, G. Bird, and J. Han. 2004. "The Amygdala and Persistent Pain." *The Neuroscientist* 10 (3): 221-234.

Newcomer, J., G. Selke, A. Melson, T. Hershey, S. Craft, K. Richards, and A. Alderson. 1999. "Decreased Memory Performance in Healthy Humans Induced by Stress-Level Cortisol Treatment." *Archives of General Psychiatry* 56 (6): 527-533.

Nichols, D. 2018. "N,N-dimethyltryptamine and the Pineal Gland: Separating Fact From Myth." *Journal of Psychopharmacology* 32 (1): 30-36.

O'Keefe, J., H. Abuissa, A. Sastre, D. Steinhaus, and W. Harris. 2006. "Effects of Omega-3 Fatty Acids on Resting Heart Rate, Heart Rate Recovery After Exercise, and Heart Rate Variability in Men With Healed Myocardial Infarctions and Depressed Ejection Fractions." *American Journal of Cardiology* 97 (8): 1127-1130.

Pascoe, M., and I. Bauer. 2015. "A Systematic Review of Randomized Control Trials on the Effects of Yoga on Stress Measures and Mood." *Journal of Psychiatric Research* 68:270-282.

Passani, M., P. Panula, and J. Lin. 2014. "Histamine in the Brain." *Frontiers in Systems Neuroscience* 8:64.

Paulson, O., S. Strandgaard, and L. Edvinsson. 1990. "Cerebral Autoregulation." *Cerebrovascular and Brain Metabolism Reviews* 2 (2): 161-192.

Polatin, P., R. Kinney, R. Gatchel, E. Lillo, and T. Mayer. 1993. "Psychiatric Illness and Chronic Low-Back Pain. The Mind and Spine: Which Goes First?" *Spine* 18:66-71.

Porges, S. 2011. *The Polyvagal Theory: Neurophysiological Foundations of Emotions, Attachment, Communication, and Self-Regulation.* New York: Norton.

Prince, M., A. Wimo, M. Guerchet, G. Ali, Y. Wu, and M. Prina. 2015. *World Alzheimer Report 2015—The Global Impact of Dementia: An Analysis of Prevalence, Incidence, Cost and Trends.* London: Alzheimer's Disease International (ADI).

Rao, M., and M. Gershon. 2016. "The Bowel and Beyond: The Enteric Nervous System in Neurological Disorders." *Nature Reviews Gastroenterology & Hepatology* 13:517-528.

Sakka, L., G. Coll, and J. Chazal. 2011. "Anatomy and Physiology of Cerebrospinal Fluid." *European Annals of Otorhinolaryngology, Head and Neck Diseases* 128 (6): 309-316.

Sarhad Hasan, M., M. Haydary, and F. Gandomi. 2020. "The Effect of Eight Weeks Yoga Training on the Mental Fatigue Control and Changed Balance and Knee Proprioception in Amateur Athletes: A Semi-Experimental Study." *Journal of Sport Biomechanics* 5 (4) 228-39.

Schmidt, N., J. Richey, J. Zvolensky, and J. Maner. 2008. "Exploring Human Freeze Responses to a Threat Stressor." *Journal of Behavior Therapy and Experimental Psychiatry* 39 (3): 292-304.

Seres, J. 2003. "Evaluating the Complex Chronic Pain Patient." *Neurosurgery Clinics of North America* 14:339-352.

Sharma, M., and T. Haider. 2013. "Yoga as an Alternative and Complementary Therapy for Patients Suffering From Anxiety: A Systematic Review." *Journal of Evidence-Based Complementary & Alternative Medicine* 18 (1): 15-22.

Sherrington, C. 1906. *The Integrative Action of the Nervous System.* New York: C. Scribner's Sons.

Sherwood, C., C. Stimpson, M. Raghanti, D. Wildman, M. Uddin, L. Grossman, M. Goodman, J. Redmond, C. Bonar, J. Erwin, and P. Hof. 2006. "Evolution of Increased Glia–Neuron Ratios in the Human Frontal Cortex." *Proceedings of the National Academy of Sciences of the United States of America* 103:13606-13611.

Spector, R., R. Snodgrass, and C. Johanson. 2015. "A Balanced View of the Cerebrospinal Fluid Composition and Functions: Focus on Adult Humans." *Experimental Neurology* 273:57-68.

Streeter, C., T. Whitfield, L. Owen, et al. 2010. "Effects of Yoga Versus Walking on Mood, Anxiety, and Brain GABA Levels: A Randomized Controlled MRS Study." *Journal of Alternative and Complementary Medicine* 16 (11): 1145-1152.

Sujan, M., K. Deepika, S. Mulakur, A. John, and T. Sathyaprabha. 2015. "Effect of Bhramari Pranayama (Humming Bee Breath) on Heart Rate Variability and Hemodynamic—A Pilot Study." *Autonomic Neuroscience* 192 (82): 1.

Tan, D., B. Xu, X. Zhou, and R. Reiter. 2018. "Pineal Calcification, Melatonin Production, Aging, Associated Health Consequences and Rejuvenation of the Pineal Gland." *Molecules* 23 (2): 301.

Tolahunase, M., R. Sagar, M. Faiq, and R. Dada. 2018. "Yoga- and Meditation-Based Lifestyle Intervention Increases Neuroplasticity and Reduces Severity of Major Depressive Disorder: A Randomized Controlled Trial." *Restorative Neurology and Neuroscience* 36 (3): 423-442.

Tysnes, O., and A. Storstein. 2017. "Epidemiology of Parkinson's Disease." *Journal of Neural Transmission* 124:901-905.

Walker, M., and R. Stickgold. 2004. "Sleep-Dependent Learning and Memory Consolidation." *Neuron* 4 (1): 121-133.

Watkins, P., K. Woodward, T. Stone, and R. Kolts. 2003. "Gratitude and Happiness: Development of a Measure of Gratitude and Relationships With Subjective Well-Being." *Social Behavior and Personality: An International Journal* 31 (5): 431-452.

Wood, A., J. Froh, and A. Geraghty. 2010. "Gratitude and Well-Being: A Review and Theoretical Integration." *Clinical Psychology Review* 30 (7): 890-905.

World Health Organization. 2012. "Depression: A Global Crisis." www.who.int/mental_health/management/depression/wfmh_paper_depression_wmhd_2012.pdf?Ua=1.

World Health Organization. 2016. *International Statistical Classification of Diseases and Related Health Problems.* 10th rev.Geneva, Switzerland: WHO.

Zhou, Y., and N. Danbolt. 2014. "Glutamate as a Neurotransmitter in the Healthy Brain." *Journal of Neural Transmission* 121 (8): 799-817.

Chapter 3

Aliverti, A. 2016. "The Respiratory Muscles During Exercise." *Breathe (Sheff)* 12 (2): 165-168. https://doi.org/10.1183/20734735.008116.

Ball, M.J., and J. Rahilly. 2003. *Phonetics: The Science of Speech.* London: Arnold.

Barker, N., and M.L. Everard. 2015. "Getting to Grips With 'Dysfunctional Breathing'." *Paediatric Respiratory Reviews* 16 (1): 53-61.

Barker, N.J., M. Jones, N.E. O'Connell, and M.L. Everard. 2013. "Breathing Exercises for Dysfunctional Breathing/Hyperventilation Syndrome in Children." *Cochrane Database of Systematic Reviews* 12.

Barr, K.P., M. Griggs, and T. Cadby. 2005. Lumbar Stabilization: Core Concepts and Current Literature, Part 1. *American Journal of Physical Medicine and Rehabilitation* 84:473-480.

Bellemare, F., A. Jeanneret, and J. Couture. 2003. "Sex Differences in Thoracic Dimensions and Configuration." *American Journal of Respiratory and Critical Care Medicine* 168 (3): 305-312. https://doi.org/10.1164/rccm.200208-876OC.

Bordoni, B., and E. Zanier. 2013. Anatomic Connections of the Diaphragm: Influence of Respiration on the Body System. *Journal of Multidisciplinary Healthcare* 6 (July 25): 281-291. https://doi.org/10.2147/JMDH.S45443.

Cacioppo, J.T., G.G. Bernston, J.T. Larsen, K.M. Poehlmann, and T. Ito. 2000. *The Psychophysiology of Emotion.* 2nd ed. New York: Guilford Press.

Carim-Todd, L., S.H. Mitchell, and B.S. Oken. 2013. "Mind–Body Practices: An Alternative, Drug-Free Treatment for Smoking Cessation? A Systematic Review of the Literature." *Drug and Alcohol Dependence* 132 (3): 399-410. https://doi.org/10.1016/j.drugalcdep.2013.04.014.

Clark, B. 2018. *Your Spine, Your Yoga: Developing Stability and Mobility for Your Spine.* Vancouver: Wild Strawberry Productions.

Cottle, M.H., R.M. Loring, G.G. Fischer, and I.E. Gaynon. 1958. "The Maxilla-Premaxilla Approach to Extensive Nasal Septum Surgery." *AMA Archives of Otolaryngology* 68 (3): 301-313.

Cramer, H., P. Posadzki, G. Dobos, and J. Langhorst. 2014. "Yoga for Asthma: A Systematic Review and Meta-Analysis." *Annals of Allergy, Asthma, & Immunology* 112 (6): 503-510.e5. https://doi.org/10.1016/j.anai.2014.03.014.

Cresswell, A.G., L. Oddsson, and A. Thorstensson. 1994. "The Influence of Sudden Perturbations on Trunk Muscle Activity and Intraabdominal Pressure While Standing." *Experimental Brain Research* 98:336-341.

De Couck, M., R. Caers, L. Musch, J. Fliegauf, A. Giangreco, and Y. Gidron. 2019. "How Breathing Can Help You Make Better Decisions: Two Studies on the Effects of Breathing Patterns on Heart Rate Variability and Decision-Making in Business Cases." *International Journal of Psychophysiology* 139:1-9.

Dimitriadis, Z., E. Kapreli, N. Strimpakos, and J. Oldham. 2014. "Pulmonary Function of Patients With Chronic Neck Pain: A Spirometry Study." *Respiratory Care* 59 (4): 543-549.

Ebenbichler, G.R., L.I. Oddsson, J. Kollmitzer, et al. 2001. "Sensory-Motor Control of the Lower Back: Implications for Rehabilitation. *Medicine and Science in Sports and Exercise* 33:1889-1898.

Estenne, M., C. Pinet, and A. De Troyer. 2000. "Abdominal Muscle Strength in Patients With Tetraplegia." *American Journal of Respiratory and Critical Care Medicine* 161 (3): 707-712.

Femina. 2020. "Health Benefits of Kapalbhati Pranayam in Yoga." Femina. www.femina.in/wellness/fitness/health-benefits-of-kapalbhati-pranayam-in-yoga-142649.html.

Flenady, T., T. Dwyer, and J. Applegarth. 2017. "Accurate Respiratory Rates Count: So Should You!" *Australasian Emergency Nursing Journal* 20 (1): 45-47.

Gardner, W.N. 1996. "The Pathophysiology of Hyperventilation Disorders." *Chest* 109 (2): 516-534.

Gerritsen, R.J.S., and G.P.H. Band. 2018. "Breath of Life: The Respiratory Vagal Stimulation Model of Contemplative Activity." *Frontiers in Human Neuroscience* 12:397.

Han, J.N., G. Gayan-Ramirez, R. Dekhuijzen, and M. Decramer. 1993. "Respiratory Function of the Rib Cage Muscles." *European Respiratory Journal* 6 (5): 722-728.

Hodges, P. 2004. "Abdominal Mechanism and Support of the Lumbar Spine and Pelvis." In *Therapeutic Exercise for Lumbopelvic Stabilization*, 2nd ed., edited by C. Richardson, 31-58. Edinburgh, Churchill Livingstone.

Holmes, S.W., R. Morris, P.R. Clance, and R.T. Putney. 1996. "Holotropic Breathwork: An Experiential Approach to Psychotherapy." *Psychotherapy: Theory, Research, Practice, Training* 33 (1): 114-120.

Jones, M., A. Harvey, L. Marston, and N.E. O'Connell. 2013. "Breathing Exercises for Dysfunctional Breathing/Hyperventilation Syndrome in Adults." *Cochrane Database of Systematic Reviews* 5.

Jung, S.I., N.K. Lee, K.W. Kang, K. Kim, and D.Y. Lee. 2016. The Effect of Smartphone Usage Time on Posture and Respiratory Function. *Journal of Physical Therapy Science* 28 (1): 186-189. https://doi.org/10.1589/jpts.28.186.

Kaminoff, L., and A. Matthews. 2007. *Yoga Anatomy.* Champaign, IL: Human Kinetics.

Kreibig, S.D. 2010. "Autonomic Nervous System Activity in Emotion: A Review." *Biological Psychology* 84 (3): 394-421. https://doi.org/10.1016/j.biopsycho.2010.03.010.

Leung, Richard ST, John S. Floras, and T. Douglas Bradley. "Respiratory modulation of the autonomic nervous system during Cheyne–Stokes respiration." *Canadian Journal of Physiology and Pharmacology* 84, no. 1 (2006): 61-66.

McKeown, P., C. O'Connor-Reina, and G. Plaza. 2021. "Breathing Re-Education and Phenotypes of Sleep Apnea: A Review." *Journal of Clinical Medicine* 10 (3): 471.

Milanesi, R., and R.C. Caregnato. 2016. "Intra-Abdominal Pressure: An Integrative Review." *Einstein (São Paulo)* 14 (3): 423-430. https://doi.org/10.1590/S1679-45082016RW3088.

Nestor, J. 2020. *Breath: The New Science of a Lost Art.* London: Penguin UK.

Paschall, J. 2013. "5 Reasons to Practice Breath of Fire Yoga Gaia." Gaia. www.gaia.com/article/5-reasons-practice-breath-fire-yoga.

Pradhan, B. 2015. "Yoga Can Spur 10% Growth and 'Cure' Homosexuals, Says Baba Ramdev." *Mint.* www.livemint.com/Politics/BjkzSSfs9SwbMxyBblkJLP/Yoga-can-spur-10-growth-and-cure-homosexuals-says-Baba-R.html.

Santino, T.A., G.S. Chaves, D.A. Freitas, G.A. Fregonezi, and K.M. Mendonça. 2020. "Breathing Exercises for Adults With Asthma." *Cochrane Database Syst Reviews* 3 (3): CD001277. https://doi.org/10.1002/14651858.CD001277.pub4.

Seals, D.R., N.O. Suwarno, and J.A. Dempsey. 1990. "Influence of Lung Volume on Sympathetic Nerve Discharge in Normal Humans." *Circulation Research* 67 (1): 130-141.

Sikter, A., E. Frecska, I.M. Braun, X. Gonda, and Z. Rihmer. 2007. "The Role of Hyperventilation: Hypocapnia in the Pathomechanism of Panic Disorder." *Brazilian Journal of Psychiatry* 29 (4): 375-379.

Singleton, M. 2010. *Yoga Body: The Origins of Modern Posture Practice.* Oxford: Oxford University Press.

Sovik, R. (n.d.). *Learn Kapalabhati (Skull Shining Breath).* Yoga International. https://yogainternational.com/article/view/learn-kapalabhati-skull-shining-breath

Strøm-Tejsen, P., D. Zukowska, P. Wargocki, and D.P. Wyon. 2016. "The Effects of Bedroom Air Quality on Sleep and Next-Day Performance." *Indoor Air* 26:679-686. https://doi.org/10.1111/ina.12254.

Szczygieł, E., K. Węglarz, K. Piotrowski, T. Mazur, S. Mętel, and J. Golec. 2015. Biomechanical Influences on Head Posture and the Respiratory Movements of the Chest. *Acta of Bioengineering and Biomechanics* 17 (2): 143-148.

Talasz, H., Kremser, C., Kofler, M., Kalchschmid, E., Lechleitner, M., and Rudisch, A. 2011. "Phase-Locked Parallel Movement of Diaphragm and Pelvic Floor During Breathing and Coughing—A Dynamic MRI Investigation in Healthy Females." *International Urogynecology Journal* 22 (1): 61-68. https://doi.org/10.1007/s00192-010-1240-z.

Vidotto, L.S., C.R.F., Carvalho, A. Harvey, and M. Jones. 2019. "Dysfunctional Breathing: What Do We Know?" *Journal of Brazilian Pneumology* 45 (1): e20170347. https://doi.org/10.1590/1806-3713/e20170347.

Vostatek, P., D. Novák, T. Rychnovský, and S. Rychnovská. 2013. "Diaphragm Postural Function Analysis Using Magnetic Resonance Imaging." *PLOS ONE* 8 (3): e56724. https://doi.org/10.1371/journal.pone.0056724.

Wikipedia. 2020. "Ujjayi Breath." https://en.wikipedia.org/wiki/Ujjayi_breath.

Wilson, J. 2009. "Hindu Guru Claims Homosexuality Can Be 'Cured' by Yoga." *The Telegraph.* www.telegraph.co.uk/news/worldnews/asia/india/5780028/Hindu-guru-claims-homosexuality-can-be-cured-by-yoga.html.

Wim Hof Method. 2020. *The Benefits of Breathing Exercises Wim Hof Method.* www.wimhofmethod.com/breathing-exercises.

Wu, L.L., Z.K. Lin, H.D. Weng, Q.F. Qi, J. Lu, and K.X. Liu. 2018. "Effectiveness of Meditative Movement on COPD: A Systematic Review and Meta-Analysis." *International Journal of Chronic Obstructive Pulmonary Disease* 13 (April 17): 1239-1250. https://doi.org/10.2147/COPD.S159042.

Yang, Z.Y., H.B. Zhong, C. Mao, et al. 2016. Yoga for Asthma. *Sao Paulo Medical Journal* 134 (4): 368. https://doi.org/10.1590/1516-3180.20161344T2.

Zaccaro, A., A. Piarulli, M. Laurino, et al. 2018. "How Breath-Control Can Change Your Life: A Systematic Review on Psycho-Physiological Correlates of Slow Breathing." *Frontiers in Human Neuroscience* 12 (September 7): 353. https://doi.org/10.3389/fnhum.2018.00353.

Chapter 4

Aird, W. 2011. "Discovery of the Cardiovascular System: From Galen to William Harvey." *Journal of Thrombosis and Haemostasis* 9 (Suppl. 1): 118-129.

Al-Khazraji, B., and J. Shoemaker. 2018. "The Human Cortical Autonomic Network and Volitional Exercise in Health and Disease." *Applied Physiology, Nutrition, and Metabolism* 43 (11): 1122-1130.

Baker, L. 2019. "Physiology of Sweat Gland Function: The Roles of Sweating and Sweat Composition in Human Health." *Temperature (Austin)* 6 (3): 211-259.

Bates, S., and J. Ginsberg. 2001. "Pregnancy and Deep Vein Thrombosis." *Seminars in Vascular Medicine* 1 (1): 97-104.

Bernardia, L., A. Gabuttia, C. Portaa, and L. Spicuzza. 2001. "Slow Breathing Reduces Chemoreflex Response to Hypoxia and Hypercapnia and Increases Baroreflex Sensitivity." *Journal of Hypertension* 19:2221-2229.

Bhavanani, A. 2016. "Yoga and Cardiovascular Health: Exploring Possible Benefits and Postulated Mechanisms." *SM Journal of Cardiovascular Diseases* 1 (1): 1003.

Boyett, M., A. D'Souza, H. Zhang, G. Morris, H. Dobrzynski, and O. Monfredi. 2013. "Viewpoint: Is the Resting Bradycardia in Athletes the Result of Remodeling of the Sinoatrial Node Rather Than High Vagal Tone?" *Journal of Applied Physiology* 114 (9): 1351-1355.

Buccelletti, E., E. Gilardi, E. Scaini, L. Galiuto, R. Persiani, A. Biondi, F. Basile, and N. Silveri. 2009. "Heart Rate Variability and Myocardial Infarction: Systematic Literature Review and Metanalysis." *European Review for Medical Pharmacological Sciences* 13 (4): 299-307.

Byeon, K., J. Choi, J. Yang, J. Sung, S. Park, J. Oh, and K. Hong. 2012. "The Response of the Vena Cava to Abdominal Breathing." *Journal of Alternative and Complementary Medicine* 18 (2): 153-157.

Cannon, W. 1932. *The Wisdom of the Body.* New York: W.W. Norton & Company.

Cole, R. 1989. "Postural Baroreflex Stimuli May Affect EEG Arousal and Sleep in Humans." *Journal of Applied Physiology* 67 (6): 2369-2375.

Cooney, M., E. Vartiainen, T. Laatikainen, A. Juolevi, A. Dudina, and I. Graham. 2010. "Elevated Resting Heart Rate Is an Independent Risk Factor for Cardiovascular Disease in Healthy Men and Women." *American Heart Journal* 159 (4): 612-619.

Cramer, H., R. Lauche, H. Haller, N. Steckhan, A. Michalsen, and G. Dobos. 2014. "Effects of Yoga on Cardiovascular Disease Risk Factors: A Systematic Review and Meta-Analysis." *International Journal of Cardiology* 173 (2): 170-183.

Dick, T., J. Mims, Y. Hsieh, K. Morris, and E. Wehrwein. 2014. "Increased Cardio-Respiratory Coupling Evoked by Slow Deep Breathing Can Persist in Normal Humans." *Respiratory Physiology and Neurobiology* 204:99-111.

Etulain, J. 2018. "Platelets in Wound Healing and Regenerative Medicine." *Platelets* 29 (6): 556-568.

Evans, C., F. Fowkes, C. Ruckley, and A. Lee. 1999. "Prevalence of Varicose Veins and Chronic Venous Insufficiency in Men and Women in the General Population: Edinburgh Vein Study." *Journal of Epidemiology and Community Health* 53:149-153.

Fagard, R. 2003. "Athlete's Heart." *Heart* 89 (12): 1455-1461.

Fryar, C., T-C. Chen, and X. Li. 2012. *Prevalence of Uncontrolled Risk Factors for Cardiovascular Disease: United States, 1999-2010.* Hyattsville, MD: National Center for Health Statistics. www.cdc.gov/nchs/data/databriefs/db103.pdf.

Gavish, I., and B. Brenner. 2011. "Air Travel and the Risk of Thromboembolism." *Internal and Emergency Medicine* 6 (2): 113-116.

Gillespie, C., E. Kuklina, P. Briss, N. Blair, and Y. Hong. 2011. "Vital Signs: Prevalence, Treatment, and Control of Hypertension, United States, 1999-2002 and 2005-2008." *Morbidity and Mortality Weekly Report* 60 (4): 103-108.

Haennel, R., K. Teo, G. Snydmiller, H. Quinney, and C. Kappagoda. 1988. "Short-Term Cardiovascular Adaptations to Vertical Head-Down Suspension." *Archives of Physical Medicine and Rehabilitation* 69 (5): 352-357.

Hagins, M., R. States, T. Selfe, and K. Innes. 2013. "Effectiveness of Yoga for Hypertension: Systematic Review and Meta-Analysis." *Evidence-Based Complementary and Alternative Medicine* 2013: 649836.

Hall, J. 2015. *Guyton and Hall Textbook of Medical Physiology.* 13th ed. Philadelphia: Saunders.

Heron, M. 2019. "Deaths: Leading Causes for 2017." *National Vital Statistics Reports* 68 (6): 1-77.

Higgins, J. 2015. "Red Blood Cell Population Dynamics." *Clinics in Laboratory Medicine* 35 (1): 43-57.

Jensen-Urstad, K., B. Saltin, M. Ericson, N. Storck, and M. Jensen-Urstad. 1997. "Pronounced Resting Bradycardia in Male Elite Runners Is Associated With High Heart Rate Variability." *Scandinavian Journal of Medicine and Science in Sports* 7:274-278.

Joseph, C., C. Porta, G. Casucci, N. Casiraghi, M. Maffeis, M. Rossi, and L. Bernardi. 2005. "Slow Breathing Improves Arterial Baroreflex Sensitivity and Decreases Blood Pressure in Essential Hypertension." *Hypertension* 46 (4): 714-718.

Kearney, P., M. Whelton, K. Reynolds, P. Muntner, P. Whelton, and J. He. 2005. "Global Burden of Hypertension: Analysis of Worldwide Data." *Lancet* 365 (9455): 217-223.

Keenan, C., and R. White. 2007. "The Effects of Race/Ethnicity and Sex on the Risk of Venous Thromboembolism." *Current Opinion in Pulmonary Medicine* 13 (5): 377-383.

Kravtsov, P., S. Katorkin, V. Volkovoy, and Y. Sizonenko. 2016. "The Influence of the Training of the Muscular Component of the Musculo-Venous Pump in the Lower Extremities on the Clinical Course of Varicose Vein Disease." *Vopr Kurortol Fizioter Lech Fiz Kult* 93 (60): 33-36.

Lasater, J. 2017. "Compassionate Dying." *Yoga Journal.* Last modified April 5, 2017. www.yogajournal.com/yoga-101/compassionate-dying

Lubitz, S. 2004. "Early Reactions to Harvey's Circulation Theory: The Impact on Medicine." *Mount Sinai Journal of Medicine* 71:274-280.

Masterson, M., A. Morgan, C. Multer, and D. Cipriani. 2006. "The Role of Lower Leg Muscle Activity in Blood Pressure Maintenance of Older Adults." *Clinical Kinesiology* 60 (2): 8-17.

McEwen, B., and E. Stellar. 1993. "Stress and the Individual. Mechanisms Leading to Disease." *Archives of Internal Medicine* 153 (18): 2093-2101.

Meneton, P., X. Jeunemaitre, H. de Wardener, and G. MacGregor. 2005. "Links Between Dietary Salt Intake, Renal Salt Handling, Blood Pressure, and Cardiovascular Diseases." *Physiology Reviews* 85 (2): 679-715.

Messerli, F., B. Williams, and E. Ritz. 2007. "Essential Hypertension." *Lancet* 370 (9587): 591-603.

Miller, J., D. Pegelow, A. Jacques, and J. Dempsey. 2005. "Skeletal Muscle Pump Versus Respiratory Muscle Pump: Modulation of Venous Return From the Locomotor Limb in Humans." *Journal of Physiology* 563: 925-943.

Norcliffe-Kaufmann, L., H. Kaufmann, J. Martinez, S. Katz, L. Tully, and H. Reynolds. 2016. "Autonomic Findings in Takotsubo Cardiomyopathy." *American Journal of Cardiology* 117 (2): 206-213.

Palatini, P. 2011. "Role of Elevated Heart Rate in the Development of Cardiovascular Disease in Hypertension." *Hypertension* 58:745-750.

Parshad, O., A. Richards, and M. Asnani. 2011. "Impact of Yoga on Haemodynamic Function in Healthy Medical Students." *West Indian Medical Journal* 60 (2): 148-152.

Pearce, J. 2007. "Malpighi and the Discovery of Capillaries." *European Neurology* 58(4): 253-255.

Posadzki, P., A. Kuzdzal, M. Lee, and E. Ernst 2015. "Yoga for Heart Rate Variability: A Systematic Review and Meta-Analysis of Randomized Clinical Trials." *Applied Psychophysiology and Biofeedback* 40:239-249.

Posner, M., and S. Petersen. 1990. "The Attention System of the Human Brain." *Annual Review of Neuroscience* 13:25-42.

Quer, G., P. Gouda, M. Galarnyk, E. Topol, and S. Steinhubl. 2020. "Inter- and Intraindividual Variability in Daily Resting Heart Rate and Its Associations With Age, Sex, Sleep, BMI, and Time of Year: Retrospective, Longitudinal Cohort Study of 92,457 Adults." *PLOS ONE* 15(2): e0227709.

Racinais, S., J. Alonso, A. Coutts, A. Flouris, O. Girard, J. González-Alonso, C. Hausswirth, et al. 2015. "Consensus Recommendations on Training and Competing in the Heat." *British Journal of Sports Medicine* 49:1164-1173.

Razin, A. 1977. "Upside-Down Position to Terminate Tachycardia of Wolff-Parkinson-White Syndrome." *New England Journal of Medicine* 296 (26): 1535-1536.

Reimers, A., G. Knapp, and C. Reimers. 2018. "Effects of Exercise on the Resting Heart Rate: A Systematic Review and Meta-Analysis of Interventional Studies." *Journal of Clinical Medicine* 7 (12): 503.

Roger, V., A. Go, D. Lloyd-Jones, E. Benjamin, J. Berry, W. Borden, D. Bravata, S. Dai, E. Ford, C. Fox, et al. 2012. "Executive Summary: Heart Disease and Stroke Statistics—2012 Update: A Report From the American Heart Association." *Circulation* 125 (1): 188-197.

Rosengren, A., S. Hawken, S. Ounpuu, K. Sliwa, M. Zubaid, W. Almahmeed, K. Blackett, C. Sitthiamorn, H. Sato, and S. Yusuf. 2004. "Association of Psychosocial Risk Factors With Risk of Acute Myocardial Infarction in 11119 Cases and 13648 Controls From 52 Countries (the INTERHEART Study): Case-Control Study." *Lancet* 364:953-962.

Schiweck, C., D. Piette, D. Berckmans, S. Claes, and E. Vrieze. 2019. "Heart Rate and High Frequency Heart Rate Variability During Stress as Biomarker for Clinical Depression. A Systematic Review." *Psychological Medicine* 49 (2): 200-211.

Schneider, R., F. Staggers, C. Alexander, W. Sheppard, M. Rainforth, K. Kondwani, S. Smith, and C. King. 1995. "A Randomised Controlled Trial of Stress Reduction for Hypertension in Older African Americans." *Hypertension* 26 (5): 820-857.

Schröder, J., and J. Harder. 2006. "Antimicrobial Skin Peptides and Proteins." *Cellular and Molecular Life Sciences* 63 (4): 469-486.

Selvamurthy, W., K. Sridharan, U. Ray, R. Tiwary, K. Hegde, U. Radhakrishan, and K. Sinha. 1998. "A New Physiological Approach to Control Essential Hypertension." *Indian Journal of Physiology and Pharmacology* 42 (2): 205-213.

Sharma, K., P. Thirumaleshwara, K. Udayakumar, and B. Savitha. 2014. "A Study on the Effect of Yoga Therapy on Anaemia in Women." *European Scientific Journal* 10 (21): 283-290.

Silverstein, M., J. Heit, D. Mohr, T. Petterson, W. O'Fallon, and L. Melton. 1998. "Trends in the Incidence of Deep Vein Thrombosis and Pulmonary Embolism: A 25-Year Population-Based Study." *Archives of Internal Medicine* 158 (6): 585-593.

Spreeuw, J., and I. Owadally. 2013. "Investigating the Broken-Heart Effect: A Model for Short-Term Dependence Between the Remaining Lifetimes of Joint Lives." *Annals of Actuarial Science* 7 (2): 236-257.

Summers, C., S. Rankin, A. Condliffe, N. Singh, A. Peters, and E. Chilvers. 2010. "Neutrophil Kinetics in Health and Disease." *Trends in Immunology* 31 (8): 318-324.

Tai, Y., and C. Colaco. 1981. "Upside-Down Position for Paroxysmal Supraventricular Tachycardia." *Lancet* 2 (8258): 1289.

Takahashi, H., M. Yoshika, Y. Komiyama, and M. Nishimura. 2011. "The Central Mechanism Underlying Hypertension: A Review of the Roles of Sodium Ions, Epithelial Sodium Channels, the Renin-Angiotensin-Aldosterone System, Oxidative Stress and Endogenous Digitalis in the Brain." *Hypertension Research* 34 (11): 1147-1160.

Tanner, J. 1951. "The Relationships Between the Frequency of the Heart, Oral Temperature and Rectal Temperature in Man at Rest." *Journal of Physiology* 115 (4): 391-409.

Tsuji, H., F. Venditti, E. Manders, J. Evans, M. Larson, C. Feldman, and D. Levy. 1994. "Reduced Heart Rate Variability and Mortality Risk in an Elderly Cohort. The Framingham Heart Study." *Circulation* 90 (2): 878-883.

Tyagi, A., and M. Cohen. 2016. "Yoga and Heart Rate Variability: A Comprehensive Review of the Literature." *International Journal of Yoga* 9 (2): 97=113.

Vijayalakshmi, P., Madanmohan, A.B. Bhavanani, A. Patil, and K. Babu. 2004. "Modulation of Stress Induced by Isometric Handgrip Test in Hypertensive Patients Following Yogic Relaxation Training." *Indian Journal of Physiology and Pharmacology* 48:59-64.

Watabe, A., T. Sugawara, K. Kikuchi, K. Yamasaki, S. Sakai, and S. Aiba. 2013. "Sweat Constitutes Several Natural Moisturizing Factors, Lactate, Urea, Sodium, and Potassium." *Journal of Dermatological Science* 72 (2): 177-182.

Whelton, P., R. Carey, W. Aronow, D. Casey, K. Collins, C. Dennison Himmelfarb, S. DePalma, S. Gidding, K. Jamerson, D. Jones, et al. 2018. "2017 ACC/AHA/AAPA/ABC/ACPM/AGS/APhA/ASH/ASPC/NMA/PCNA Guideline for the Prevention, Detection, Evaluation, and Management of High Blood Pressure in Adults." *Journal of the American College of Cardiology* 71 (19): e127-e248.

Williamson, J., R. McColl, and D. Mathews. 2004. "Changes in Regional Cerebral Blood Flow Distribution During Postexercise Hypotension in Humans." *Journal of Applied Physiology* 96 (2): 719-724.

Yoshizumi, M., J. Abe, K. Tsuchiya, B. Berk, and T. Tamaki. 2003. "Stress and Vascular Responses: Athero-Protective Effect of Laminar Fluid Shear Stress in Endothelial Cells: Possible and Mitogen-Activated Protein Kinases." *Journal of Pharmacological Sciences* 9: 172-176.

Chapter 5

Adair, T., D. Moffatt, A. Paulsen, and A. Guyton. 1982. "Quantitation of Changes in Lymph Protein Concentration During Lymph Node Transit." *American Journal of Physiology* 243:H351-H359.

Ader, R., and N. Cohen. 1975. "Behaviorally Conditioned Immunosuppression." *Psychosomatic Medicine* 37 (4): 333-340.

Aspelund, A., M. Robciuc, S. Karaman, T. Makinen, and K. Alitalo. 2016. "Lymphatic System in Cardiovascular Medicine." *Circulation Research* 118 (3): 515=530.

Auckland, K. 2005. "Arnold Heller and the Lymph Pump." *Acta Physiologica Scandinavica* 185:171-180.

Balasubramaniam, M., S. Telles, and P. Doraiswamy. 2013. "Yoga on Our Minds: A Systematic Review of Yoga for Neuropsychiatric Disorders." *Frontiers in Psychiatry* 3:117.

Balloux, F., and L. van Dorp. 2017. "Q&A: What Are Pathogens, and What Have They Done to and for Us?" *BMC Biology* 15 (1): 91.

Bodey, B., S. Siegel, and H. Kaiser. 2006. *Immunological Aspects of Neoplasia—The Role of the Thymus.* Berlin: Springer Science & Business Media.

Buffart, L., J. van Uffelen, I. Riphage, J. Brug, W. van Mechelen, W. Brown, and M. Chinapaw. 2012. "Physical and Psychosocial Benefits of Yoga in Cancer Patients and Survivors, A Systematic Review and Meta-Analysis of Randomized Controlled Trials." *BMC Cancer* 12: 559.

Cappuccio, F., L. D'Elia, P. Strazzullo, and M. Miller. 2010. "Sleep Duration and All-Cause Mortality: A Systematic Review and Meta-Analysis of Prospective Studies." *Sleep* 33:585-592.

Castellani, J., I. Brenner, and S. Rhind. 2003. "Cold Exposure: Human Immune Responses and Intracellular Cytokine Expression." *Medicine and Science in Sports and Exercise* 34:2013-2020.

Castelo-Branco, C., and I. Soveral. 2014. "The Immune System and Aging: A Review." *Gynecological Endocrinology* 30 (1): 16-22.

Cemal, Y., A. Pusic, and B. Mehrara. 2011. "Preventative Measures for Lymphedema: Separating Fact From Fiction." *Journal of the American College of Surgeons* 213 (4): 543-551.

Cheville, A., C. McGarvey, J. Petrek, S. Russo, M. Taylor, and S. Thiadens. 2003. "Lymphedema Management." *Seminars in Radiation Oncology* 13:290-301.

Choe, K., J. Jang, I. Park, Y. Kim, S. Ahn, D. Park, Y. Hong, K. Alitalo, G. Koh, and P. Kim. 2015. "Intravital Imaging of Intestinal Lacteals Unveils Lipid Drainage Through Contractility." *Journal of Clinical Investigation* 125:4042-4052.

Choi, I., S. Lee, and Y. Hong. 2012. "The New Era of the Lymphatic System: No Longer Secondary to the Blood Vascular System." *Cold Spring Harbor Perspectives in Medicine* 2(4): a006445.

Cohen, L., C. Warneke, R. Fouladi, M. Rodriguez, and A. Chaoul-Reich. 2004. "Psychological Adjustment and Sleep Quality in a Randomized Trial of the Effects of a Tibetan Yoga Intervention in Patients With Lymphoma." *Cancer* 100 (10): 2253-2260.

Deshmukh, S., F. Verde, P. Johnson, E. Fishman, and K. Macura. 2014. "Anatomical Variants and Pathologies of the Vermix." *Emergency Radiology* 21 (5): 543-552.

Dongaonkar, R., R. Stewart, H. Geissler, and G. Laine. 2010. "Myocardial Microvascular Permeability, Interstitial Oedema, and Compromised Cardiac Function." *Cardiovascular Research* 87 (2): 331-339.

Douglass, J., M. Immink, N. Piller, and S. Ullah. 2012. "Yoga for Women With Breast Cancer-Related Lymphoedema: A Preliminary 6-Month Study." *Journal of Lymphoedema* 7:30-38.

Dunne, E., B. Balletto, M. Donahue, M. Feulner, J. DeCosta, D. Cruess, E. Salmoirago-Blotcher, R. Wing, M. Carey, and L. Scott-Sheldon. 2019. "The Benefits of Yoga for People Living With HIV/AIDS: A Systematic Review and Meta-Analysis." *Complementary Therapies in Clinical Practice* 34:157-164.

Edwards, J., K. Williams, L. Kindblom, J. Meis-Kindblom, P. Hogendoorn, D. Hughes, R. Forsyth, D. Jackson, and N. Athanasou. 2008. "Lymphatics and Bone." *Human Pathology* 39:49-55.

El-Kadiki, A., and A. Sutton. 2005. "Role of Multivitamins and Mineral Supplements in Preventing Infections in Elderly People: Systematic Review and Meta-Analysis of Randomised Controlled Trials." *BMJ (Clinical Research ed.)* 330 (7496): 871.

Engeset, A., W. Olszewski, P. Jaeger, J. Sokolowski, and L. Theodorsen. 1977. "Twenty-Four Hour Variation in Flow and Composition of Leg Lymph in Normal Men." *Acta Physiologica Scandinavica* 99: 140-148.

Falkenberg, R., C. Eising, and M. Peters. 2018. "Yoga and Immune System Functioning: A Systematic Review of Randomized Controlled Trials." *Journal of Behavioral Medicine* 41:467-482.

Gashev, A. 2002. "Physiologic Aspects of Lymphatic Contractile Function." *Annals of the New York Academy of Sciences* 979: 178-187.

Gashev, A., and D. Zawieja, 2001. "Physiology of Human Lymphatic Contractility: A Historical Perspective." *Lymphology* 34 (3): 124-134.

Haaland, D., T. Sabljic, D. Baribeau, I. Mukovozov, and L. Hart. 2008. "Is Regular Exercise a Friend or Foe of the Aging Immune System? A Systematic Review." *Clinical Journal of Sport Medicine* 18:539-548.

Haapakoski, R., K.P. Ebmeier, H. Alenius, and M. Kivimäki. 2016. "Innate and Adaptive Immunity in the Development of Depression: An Update on Current Knowledge and Technological Advances." *Progress in Neuro-Psychopharmacology & Biological Psychiatry* 66: 63-72.

Hess, P., D. Rawnsley, Z. Jakus, Y. Yang, D. Sweet, J. Fu, B. Herzog, et al. 2014. "Platelets Mediate Lymphovenous Hemostasis to Maintain Blood-Lymphatic Separation Throughout Life." *Journal of Clinical Investigation* 124:273-284.

Huang, T., S. Tseng, C. Lin, C. Bai, C. Chen, C. Hung, C. Wu, and K. Tam. 2013. "Effects of Manual Lymphatic Drainage on Breast Cancer-Related Lymphedema: A Systematic Review and Meta-Analysis of Randomized Controlled Trials." *World Journal of Surgical Oncology* 11: 15.

Ironson, G., C. O'Cleirigh, M. Kumar, L. Kaplan, E. Balbin, C. Kelsch, M. Fletcher, and N. Schneiderman. 2015. "Psychosocial and Neurohormonal Predictors of HIV Disease Progression (CD4 Cells and Viral Load): A 4 Year Prospective Study." *AIDS and Behavior* 19 (8): 1388-1397.

Irwin, M., R. Olmstead, and J. Carroll. 2016. "Sleep Disturbance, Sleep Duration, and Inflammation: A Systematic Review and Meta-Analysis of Cohort Studies and Experimental Sleep Deprivation." *Biological Psychiatry* 80 (1): 40-52.

Klotz, L., S. Norman, J. Vieira, M. Masters, M. Rohling, K. Dubé, S. Bollini, F. Matsuzaki, C. Carr, and P. Riley. 2015. "Cardiac Lymphatics Are Heterogeneous in Origin and Respond to Injury." *Nature* 522 (7554): 62-67.

Levick, J., and C. Michel. 2010. "Microvascular Fluid Exchange and the Revised Starling Principle." *Cardiovascular Research* 87:198-210.

Lim, H., C. Thiam, K. Yeo, R. Bisoendial, C. Hii, K. McGrath, K. Tan, A. Heather, J. Alexander, and V. Angeli. 2013. "Lymphatic Vessels Are Essential for the Removal of Cholesterol From Peripheral Tissues by SR-BI-Mediated Transport of HDL." *Cell Metabolism* 17 (5): 671-684.

Margaris, K., and R. Black. 2012. "Modelling the Lymphatic System: Challenges and Opportunities." *Journal of Royal Society Interface* 9:601-612.

Martin, S., B. Pence, and J. Woods. 2009. "Exercise and Respiratory Tract Viral Infections." *Exercise and Sport Sciences Reviews* 37 (4): 157-164.

McNeely, M., C. Peddle, J. Yurick, I. Dayes, and J. Mackey. 2011. "Conservative and Dietary Interventions for Cancer-Related Lymphedema: A Systematic Review and Meta-Analysis." *Cancer* 117: 1136-1148.

Mislin, H. 1961. "Zur Funktionsanalyse der Lymphgefässmotorik." *Revue Suisse Zoologie* 68:228-238.

Mitchell, R., S. Archer, S. Ishman, R. Rosenfeld, S. Coles, S. Finestone, N. Friedman, T. Giordano, et al. 2019. "Clinical Practice Guideline: Tonsillectomy in Children (Update)." *Otolaryngology—Head and Neck Surgery* 160 (1): S1-S42.

Morgan, N., M. Irwin, M. Chung, and C. Wang. 2014. "The Effects of Mind–Body Therapies on the Immune System: Meta-Analysis." *PLOS ONE* 9(7): e100903.

Palmer, S., L. Albergante, C. Blackburn, and T. Newman 2018. "Thymic Involution and Disease Incidence." *Proceedings of the National Academy of Sciences of the United States of America* 115 (8): 1883-1888.

Petersen, A., and B. Pedersen. 2005. "The Anti-Inflammatory Effect of Exercise." *Journal of Applied Physiology* 98:1154-1162.

Petrek, J., and M. Heelan. 1998. "Incidence of Breast Carcinoma-Related Lymphedema." *Cancer* 83:2776-2781.

Pflicke, H., and M. Sixt. 2009. "Preformed Portals Facilitate Dendritic Cell Entry Into Afferent Lymphatic Vessels." *Journal of Experimental Medicine* 206:2925-2935.

Piller, N., G. Craig, A. Leduc, and T. Ryan. 2006. "Does Breathing Have an Influence on Lymphatic Drainage?" *Journal of Lymphoedema* 1 (1): 86-88.

Ploeger, H., T. Takken, M. de Greef, and B. Timmons. 2009. "The Effects of Acute and Chronic Exercise on Inflammatory Markers in Children and Adults With a Chronic Inflammatory Disease: A Systematic Review." *Exercise Immunology Review* 15:6-41.

Randal Bollinger, R., A. Barbas, E. Bush, S. Lin, and W. Parker. 2007. "Biofilms in the Large Bowel Suggest an Apparent Function of the Human Vermiform Appendix." *Journal of Theoretical Biology* 249 (4): 826-831.

Roblin, X., C. Neut, A. Darfeuille-Michaud, and J. Colombel. 2012. "Local Appendiceal Dysbiosis: The Missing Link Between the Appendix and Ulcerative Colitis?" *Gut* 61:635-636.

Rytter, M., L. Kolte, A. Briend, H. Friis, and V. Christensen. 2014. "The Immune System in Children With Malnutrition—A Systematic Review." *PLOS ONE* 9(8): e105017.

Sanders, N., R. Bollinger, R. Lee, S. Thomas, and W. Parker. 2013. "Appendectomy and *Clostridium Difficile* Colitis: Relationships Revealed by Clinical Observations and Immunology." *World Journal of Gastroenterol* 19 (34): 5607-5614.

Schwager, S., and M. Detmar. 2019. "Inflammation and Lymphatic Function." *Frontiers in Immunology* 10:308.

Seki, H. 1979. "Lymph Flow in Human Leg." *Lymphology* 12:2-3.

Sender, R., S. Fuchs, and R. Milo. 2016. "Revised Estimates for the Number of Human and Bacteria Cells in the Body." *PLOS Biology* 14(8): e1002533.

Sharma, M., T. Haider, and A. Knowlden. 2013. "Yoga as an Alternative and Complementary Treatment for Cancer: A Systematic Review." *The Journal of Alternative and Complementary Medicine* 19 (11): 870-875.

Shurin, M. 2012. "Cancer as an Immune-Mediated Disease." *ImmunoTargets and Therapy* 1:1-6.

Song, H., F. Fang, G. Tomasson, F. Arnberg, D. Mataix-Cols, L. Fernández de la Cruz, C. Almqvist, K. Fall, and U. Valdimarsdóttir. 2018. "Association of Stress-Related Disorders With Subsequent Autoimmune Disease." *JAMA* 319 (23): 2388-2400.

Summers, C., S. Rankin, A. Condliffe, N. Singh, A. Peters, and E. Chilvers. 2010. "Neutrophil Kinetics in Health and Disease." *Trends in Immunology* 31 (8): 318-324.

Tzeng, Y., L. Kao, S. Kao, H. Lin, M. Tsai, and C. Lee. 2015. "An Appendectomy Increases the Risk of Rheumatoid Arthritis: A Five-Year Follow-Up Study." *PLOS ONE* 10(5): e0126816.

Wanchai, A., and J. Armer. 2020. "The Effects of Yoga on Breast-Cancer-Related Lymphedema: A Systematic Review." *Journal of Health Research* (April). 409-418.

Wang, F., O. Lee, F. Feng, M. Vitiello, W. Wang, H. Benson, G. Fricchione, and J. Denninger. 2015. "The Effect of Meditative Movement on Sleep Quality: A Systematic Review." *Sleep Medicine Reviews* 30: 43-52.

Wang, W., K. Chen, Y. Pan, S. Yang, and Y. Chan. 2020. "The Effect of Yoga on Sleep Quality and Insomnia in Women With Sleep Problems: A Systematic Review and Meta-Analysis." *BMC Psychiatry* 20(1): 195.

Wang, X., P. Li, C. Pan, L. Dai, Y. Wu, and Y. Deng. 2019. "The Effect of Mind–Body Therapies on Insomnia: A Systematic Review and Meta-Analysis." *Evidence-Based Complementary and Alternative Medicine* 2019: 9359807.

Watson, N., D. Buchwald, J. Delrow, W. Altemeier, M. Vitiello, A. Pack, M. Bamshad, C. Noonan, and S. Gharib. 2017. "Transcriptional Signatures of Sleep Duration Discordance in Monozygotic Twins." *Sleep* 40(1): zsw019.

WHO. n.d. "HIV/AIDS Data and Statistics." Accessed June 19, 2020. www.who.int/hiv/data/en/

Wu, S., W. Chen, C. Muo, T. Ke, C. Fang, and F. Sung. 2015. "Association Between Appendectomy and Subsequent Colorectal Cancer Development: An Asian Population Study." *PLOS ONE* 10(2): e0118411.

Xu, R., H. Rahmandad, M. Gupta, C. Digennaro, N. Ghaffarzadegan, H. Amini, and S. Jalali. 2020. "The Modest Impact of Weather and Air Pollution on COVID-19 Transmission." *Available at SSRN 3593879.*

Chapter 6

Akil, H., S.J. Watson, E. Young, M.E. Lewis, H. Khachaturian, and J.M. Walker. 1984. "Endogenous Opioids: Biology and Function." *Annual Review of Neuroscience* 7:223-255.

Altaye, K.Z., S. Mondal, K. Legesse, et al. 2019. "Effects of Aerobic Exercise on Thyroid Hormonal Change Responses Among Adolescents With Intellectual Disabilities." *BM Journal of Open Sport & Exercise Medicine* 5:e000524. doi: 10.1136/bmjsem-2019-000524.

American Psychological Association. 2018. "Stress Effects on the Body." www.apa.org/topics/stress/body.

Ayano, G. 2016. "Dopamine: Receptors, Functions, Synthesis, Pathways, Locations and Mental Disorders: Review of Literatures." *Journal of Mental Disorders and Treatment* 2 (2). https://doi.org/10.4172/2471-271X.1000120.

Bansal, A., A. Kaushik, C.M. Singh, V. Sharma, and H. Singh. 2015. "The Effect of Regular Physical Exercise on the Thyroid Function of Treated Hypothyroid Patients: An Interventional Study at a Tertiary Care Center in Bastar Region of India." *Archives of Medicine and Health Sciences* 3 (2): 244.

Benvenutti, M.J., E. da Sliva Alves, S. Michael, D. Ding, E. Stamatakis, and K.M. Edwards. 2017. "A Single Session of Hatha Yoga Improves Stress Reactivity and Recovery After an Acute Psychological Stress Task—A Counterbalanced, Randomized-Crossover Trial in Healthy Individuals." *Complementary Therapies in Medicine* 35:120-126.

Bird, S.R., and J.A. Hawley. 2017. "Update on the Effects of Physical Activity on Insulin Sensitivity in Humans." *BMJ Open Sport & Exercise Medicine* 2(1): e000143.

Brinsley, J., F. Schuch, O. Lederman, D. Girard, M. Smout, M.A. Immink, B. Stubbs, J. Firth, K. Davison, and S. Rosenbaum. 2020. "Effects of Yoga on Depressive Symptoms in People With Mental Disorders: A Systematic Review and Meta-Analysis." *British Journal of Sports Medicine* 55(17): 992-1000.

Cahill, C.A. 1989. "Beta-Endorphin Levels During Pregnancy and Labor: A Role in Pain Modulation?" *Nursing Research* 38 (4): 200-203.

Chaudhry, S.R., and W. Gossman. 2020. "Biochemistry, Endorphin." *StatPearls* [Internet https://www.ncbi.nlm.nih.gov/books/NBK470306/].

Chu, B., K. Marwaha, and D. Ayers. 2020. "Physiology, Stress Reaction." *StatPearls* [Internet https://www.ncbi.nlm.nih.gov/books/NBK541120/].

Ciloglu, F., I. Peker, A. Pehlivan, et al. 2005. "Exercise Intensity and Its Effects on Thyroid Hormones" [published correction appears in 2006 *Neuroendocrinology Letters* 27 (3): 292]. *Neuroendocrinology Letters* 26 (6): 830-834.

Cui, J., J-H. Yan, L-M. Yan, L. Pan, J-J. Le, and Y-Z. Guo. 2017. "Effects of Yoga in Adults With Type 2 Diabetes Mellitus: A Meta-Analysis." *Journal of Diabetes Investigation* 8 (2): 201-209.

Dfarhud, D., M. Malmir, and M. Khanahmadi. 2014. "Happiness & Health: The Biological Factors—Systematic Review Article." *Iran Journal of Public Health* 43 (11): 1468-1477.

Enevoldson, T. P. 2004. "Recreational drugs and their neurological consequences." *Journal of Neurology, Neurosurgery & Psychiatry* 75, no. suppl 3: iii9-iii15.

Fisher, B.E., Q. Li, A. Nacca, et al. 2013. "Treadmill Exercise Elevates Striatal Dopamine D2 Receptor Binding Potential in Patients With Early Parkinson's Disease." *Neuroreport* 24 (10): 509-514. https://doi.org/10.1097/WNR.0b013e328361dc13.

Gordon, L.A., E.Y. Morrison, D.A. McGrowder, R. Young, Y.T. Pena Fraser, E. Martorell Zamora, R.L. Alexander-Lindo, and R.R. Irving. 2008. "Effect of Exercise Therapy on Lipid Profile and Oxidative Stress Indicators in Patients With Type 2 Diabetes." *BMC Complementary and Alternative Medicine* 8 (1): 1-10.

Harvard Health Publishing. 2021. "The Lowdown on Thyroid Slowdown—Harvard Health." Harvard Health. www.health.harvard.edu/diseases-and-conditions/the-lowdown-on-thyroid-slowdown.

Herman, J.P., J.M. McKlveen, S. Ghosal, B. Kopp, A. Wulsin, R. Makinson, J. Scheimann, and B. Myers. 2011. "Regulation of the Hypothalamic-Pituitary-Adrenocortical Stress Response." *Comprehensive Physiology* 6 (2): 603-621.

International Diabetes Federation. 2019. *IDF Diabetes Atlas.* 9th ed. Brussels, Belgium: International Diabetes Federation.

Kjaer, T.W., C. Bertelsen, P. Piccini, D. Brooks, J. Alving, and H.C. Lou. 2002. "Increased Dopamine Tone During Meditation-Induced Change of Consciousness." *Brain Research. Cognitive Brain Research* 13 (2): 255-259. https://doi.org/10.1016/s0926-6410(01)00106-9.

Machin, A.J., and R.I.M. Dunbar. 2011. "The Brain Opioid Theory of Social Attachment: A Review of the Evidence." *Behaviour* 148 (9/10): 985-1025. www.jstor.org/stable/23034206.

Mak, M.K.Y., and I.S.K. Wong-Yu. 2019. "Exercise for Parkinson's Disease." *International Review of Neurobiology* 147:1-44.

Neave, N. 2008. *Hormones and Behaviour: A Psychological Approach.* Cambridge, UK: Cambridge University Press.

Pal, R., S.N. Singh, A. Chatterjee, and M. Saha. 2014. "Age-Related Changes in Cardiovascular System, Autonomic Functions, and Levels of BDNF of Healthy Active Males: Role of Yogic Practice." *Age* 36 (4): 1-17.

Pascoe, M.C., D.R. Thompson, and C.F. Ski. 2017. "Yoga, Mindfulness-Based Stress Reduction and Stress-Related Physiological Measures: A Meta-Analysis." *Psychoneuroendocrinology* 86:152-168.

Peters, R. 2006. "Ageing and the Brain." *Postgraduate Medical Journal* 82 (964): 84-88.

Petzinger, G.M., D.P. Holschneider, B.E. Fisher, S. McEwen, N. Kintz, M. Halliday, W. Toy, J.W. Walsh, J. Beeler, and M.W. Jakowec. 2015. "The Effects of Exercise on Dopamine Neurotransmission in Parkinson's Disease: Targeting Neuroplasticity to Modulate Basal Ganglia Circuitry." *Brain Plasticity* 1 (1): 29-39.

Pierce, S. 2011. "Integrative Treatment of Hypothyroidism." *Pearls for Clinicians.* University of Wisconsin Integrative Medicine. www.fammed.wisc.edu/files/webfm-uploads/documents/outreach/im/module_thyroid_clinician.pdf

Pouwer, F., N. Kupper, and M.C. Adriaanse. 2010. "Does Emotional Stress Cause Type 2 Diabetes Mellitus? A Review From the European Depression in Diabetes (EDID) Research Consortium." *Discovery Medicine* 9 (45): 112-118.

Ryff, C.D., B.H. Singer, and G. Dienberg Love. 2004. "Positive Health: Connecting Well-Being With Biology." *Philosophical Transactions of the Royal Society of London. Series B: Biological Sciences* 359 (1449): 1383-1394.

Selye, H. 1956. *The Stress of Life.* New York, NY: McGraw-Hill.

Shaw, J.E., R.A. Sicree, and P.Z. Zimmet. 2010. "Global Estimates of the Prevalence of Diabetes for 2010 and 2030." *Diabetes Research and Clinical Practice* 87 (1): 4-14.

Shuster, M. 2014. *Biology for a Changing World, With Physiology.* 2nd ed. New York: WH Freeman.

Siebers, M., S.V. Biedermann, L. Bindila, B. Lutz, and J. Fuss. 2021. "Exercise-Induced Euphoria and Anxiolysis Do Not Depend on Endogenous Opioids in Humans." *Psychoneuroendocrinology* 126:105173.

Singh, P., B. Singh, R. Dave, and R. Udainiya. 2011. "The Impact of Yoga Upon Female Patients Suffering From Hypothyroidism." *Complementary Therapies in Clinical Practice* 17 (3): 132-134.

Sprouse-Blum, A.S., G. Smith, D. Sugai, and F.D. Parsa. 2010. "Understanding Endorphins and Their Importance in Pain Management. *Hawaii Medical Journal* 69 (3): 70-71.

Vancampfort, D., J. Firth, F.B. Schuch, S. Rosenbaum, J. Mugisha, M. Hallgren, M. Probst, et al. 2017. "Sedentary Behavior and Physical Activity Levels in People With Schizophrenia, Bipolar Disorder and Major Depressive Disorder: A Global Systematic Review and Meta-Analysis." *World Psychiatry* 16 (3): 308-315.

van Galen, K.A., K.W. Ter Horst, J. Booij, S.E., la Fleur, and M.J. Serlie. 2018. "The Role of Central Dopamine and Serotonin in Human Obesity: Lessons Learned From Molecular Neuroimaging Studies." *Metabolism* 85:325-339. https://doi.org/10.1016/j.metabol.2017.09.007.

Wenzel, J.M., N.A. Rauscher, J.F. Cheer, and E.B. Oleson. 2015. "A Role for Phasic Dopamine Release Within the Nucleus Accumbens in Encoding Aversion: A Review of the Neurochemical Literature." *ACS Chemical Neuroscience* 6 (1): 16-26. https://doi.org/10.1021/cn500255p.

Chapter 7

Aldabe, D., D. Ribeiro, S. Milosavljevic, and M. Dawn Bussey. 2012. "Pregnancy-Related Pelvic Girdle Pain and Its Relationship With Relaxin Levels During Pregnancy: A Systematic Review." *European Spine Journal* 21 (9): 1769-1776.

American Academy of Pediatrics. 2012. "Breastfeeding and the Use of Human Milk." *Pediatrics* 129 (3): e827-e884.

Ashtanga Yoga Center. n.d. Web site. www.ashtangayogacenter.com/moon-days.

Baber, R. 2014. "East Is East and West Is West: Perspectives on the Menopause in Asia and the West." *Climacteric* 17 (1): 23-28.

Beddoe, A., K. Lee, S. Weiss, H. Kennedy, and C. Yang. 2010. "Effects of Mindful Yoga on Sleep in Pregnant Women: A Pilot Study." *Biological Research for Nursing* 11 (4): 363-370.

Binkley, S. 1992. "Wrist Activity in a Woman: Daily, Weekly, Menstrual, Lunar, Annual Cycles?" *Physiology and Behavior* 52:411-421.

Bulletti, C., D. de Ziegler, V. Polli, L. Diotallevi, E. Del Ferro, and C. Flamigni. 2000. "Uterine Contractility During the Menstrual Cycle." *Human Reproduction* 15 (S1): 81-89.

Buttner, M., R. Brock, M. O'Hara, and S. Stuart. 2015. "Efficacy of Yoga for Depressed Postpartum Women: A Randomized Controlled Trial." *Complementary Therapies in Clinical Practice* 21 (2): 94-100.

Cajochen, C., S. Altanay-Ekici, M. Münch, S. Frey, V. Knoblauch, and A. Wirz-Justice. 2013. "Evidence That the Lunar Cycle Influences Human Sleep." *Current Biology* 23 (15): 1485-1488.

Carvalho, M., L. Lima, C. de Lira Terceiro, D. Pinto, M. Silva, G. Cozer, and T. Couceiro. 2017. "Low Back Pain During Pregnancy." *Revista Brasileira de Anestesiologia* 67 (3): 266-270.

Cherni, Y., D. Desseauve, A. Decatoire, N. Veit-Rubinc, M. Begon, F. Pierre, and L. Fradet. 2019. "Evaluation of Ligament Laxity During Pregnancy." *Journal of Gynecology Obstetrics and Human Reproduction* 48 (5): 351-357.

Chumlea, W., C. Schubert, A. Roche, H. Kulin, P. Lee, J. Himes, and S. Sun. 2003. "Age at Menarche and Racial Comparisons in U.S. Girls." *Pediatrics* 111 (1): 110-113.

Clue. 2019. "The Myth of Moon Phases and Menstruation." Updated April 16, 2019. https://helloclue.com/articles/cycle-a-z/myth-moon-phases-menstruation.

Cramer, H., R. Lauche, J. Langhorst, and G. Dobos. 2012. "Effectiveness of Yoga for Menopausal Symptoms: A Systematic Review and Meta-Analysis of Randomized Controlled Trials." *Evidence-Based Complementary and Alternative Medicine* 2012: 863905.

Cramer, H., W. Peng, and R. Lauche. 2018. "Yoga for Menopausal Symptoms—A Systematic Review and Meta-Analysis." *Maturitas* 109:13-25.

Culver, R., J. Rotton, and I. Kelly. 1988. "Moon Mechanisms and Myths: A Critical Appraisal of Explanations of Purported Lunar Effects on Human Behavior." *Psychological Reports* 62:683-710.

Curtis, K., A. Weinrib, and J. Katz. 2012. "Systematic Review of Yoga for Pregnant Women: Current Status and Future Directions." *Evidence-Based Complementary and Alternative Medicine* 2012: 715942.

Davenport, M., A. Kathol, M. Mottola, R. Skow, V. Mcah, V. Poitras, A. Garcia, et al. 2019. "Prenatal Exercise Is Not Associated With Fetal Mortality: A Systematic Review and Meta-Analysis." *British Journal of Sports Medicine* 53:108-115.

Dawood, M. 2006. "Primary Dysmenorrhea: Advances in Pathogenesis and Management." *Obstetrics & Gynecology* 108 (2): 428-441.

Dehghan, F., B. Haerian, S. Muniandy, A. Yusof, J. Dragoo, and N. Salleh. 2014. "The Effect of Relaxin on the Musculoskeletal System." *Scandinavian Journal of Medicine and Science in Sports* 24 (4): e220-e229.

Dhawan, V., M. Kumar, P. Chaurasia, and R. Dada. 2019. "Mind–Body Interventions Significantly Decrease Oxidative DNA Damage in Sperm Genome: Clinical Implications." *Reactive Oxygen Species* 7 (19): 1-9.

Dhawan, V., M. Kumar, D. Deka, N. Malhotra, V. Dadhwal, N. Singh, and R. Dada. 2018. "Meditation & Yoga: Impact on Oxidative DNA Damage & Dysregulated Sperm Transcripts in Male Partners of Couples With Recurrent Pregnancy Loss." *The Indian Journal of Medical Research* 148:S134-S139.

Diamond, J. 2004. *The Irritable Male Syndrome: Understanding and Managing the 4 Key Causes of Depression and Aggression.* Berkeley, CA: Potter/Ten Speed/Harmony/Rodale.

Duong, H., S. Shahrukh Hashmi, T. Ramadhani, M. Canfield, A. Scheuerle, and D. Waller. 2011. National Birth Defects Prevention Study. "Maternal Use of Hot Tub and Major Structural Birth Defects." *Birth Defects Research Part A: Clinical and Molecular Teratology* 91:836-841.

Foster, R., and T. Roenneberg. 2008. "Human Responses to the Geophysical Daily, Annual and Lunar Cycles." *Current Biology* 18:R784-R794.

Gadsby, R., A. Barnie-Adshead, and C. Jagger. 1993. "A Prospective Study of Nausea and Vomiting During Pregnancy." *British Journal of General Practice* 43:245-248.

Gaitzsch, H., J. Benard, J. Hugon-Rodin, L. Benzakour, and I. Streuli. 2020. "The Effect of Mind–Body Interventions on Psychological and Pregnancy Outcomes in Infertile Women: A Systematic Review." *Archives of Women's Mental Health* 23:479-491.

Gautam, S., B. Chawla, S. Bisht, M. Tolahunase, and R. Dada. 2018. "Impact of Mindfulness Based Stress Reduction on Sperm DNA Damage." *Journal of The Anatomical Society of India* 67:124-129.

Gavin, N., K. Gaynes, B., Lohr, S. Meltzer-Brody, G. Gartlehner, and T. Swinson. 2005. "Perinatal Depression: A Systematic Review of Prevalence and incidence." *Obstetrics & Gynecology* 106:1071-1083.

Ghaffarilaleh, G., V. Ghaffarilaleh, Z. Sanamno, and M. Kamalifard. 2019. "Yoga Positively Affected Depression and Blood Pressure in Women With Premenstrual Syndrome in a Randomized Controlled Clinical Trial." *Complementary Therapies in Clinical Practice* 34:87-92.

Ghaffarilaleh, G., V. Ghaffarilaleh, Z. Sanamno, M. Kamalifard, and L. Alibaf. 2019. "Effects of Yoga on Quality of Sleep of Women With Premenstrual Syndrome." *Alternative Therapies in Health and Medicine* 25 (5): 40-47.

Goldsmith, L., and G. Weiss. 2009. "Relaxin in Human Pregnancy." *Annals of the New York Academy of Sciences* 1160:130-135.

Gong, H., C. Ni, X. Shen, T. Wu, and C. Jiang. 2015. "Yoga for Prenatal Depression: A Systematic Review and Meta-Analysis." *BMC Psychiatry* 15: 14.

Gracia, C., M. Sammel, E. Freeman, H. Lin, E. Langan, S. Kapoor, and D. Nelson. 2005. "Defining Menopause Status: Creation of a New Definition to Identify the Early Changes of the Menopausal Transition." *Menopause* 12 (2): 128-135.

Iacovides, S., I. Avidon, and F. Baker. 2015. "What We Know About Primary Dysmenorrhea Today: A Critical Review." *Human Reproduction Update* 21 (6): 762-778.

Innes, K., T. Selfe, and A. Vishnu. 2010. "Mind–Body Therapies for Menopausal Symptoms: A Systematic Review." *Maturitas* 66 (2): 135-149.

Ji, M., and Q. Yu. 2015. "Primary Osteoporosis in Postmenopausal Women." *Chronic Diseases and Translational Medicine* 1 (1): 9-13.

Ju, H., M. Jones, and G. Mishra. 2014. "The Prevalence and Risk Factors of Dysmenorrhea." *Epidemiologic Reviews* 36:104-113.

Kamalifard, M., A. Yavari, M. Asghari-Jafarabadi, G. Ghaffarilaleh, and A. Kasb-Khah. 2017. "The Effect of Yoga on Women's Premenstrual Syndrome: A Randomized Controlled Clinical Trial." *International Journal of Women's Health and Reproduction Sciences* 5:205-211.

Kanojia, S., V. Sharma, A. Gandhi, R. Kapoor, A. Kukreja, and S. Subramanian. 2013. "Effect of Yoga on Autonomic Functions and Psychological Status During Both Phases of Menstrual Cycle in Young Healthy Females." *Journal of Clinical and Diagnostic Research* 7 (10): 2133-2139.

Kim, S-D. 2019. "Yoga for Menstrual Pain in Primary Dysmenorrhea: A Meta-Analysis of Randomized Controlled Trials." *Complementary Therapies in Clinical Practice* 36: 94-99.

Kinser, P., J. Pauli, N. Jallo, M. Shall, K. Karst, M. Hoekstra, and A. Starkweather. 2017. "Physical Activity and Yoga-Based Approaches for Pregnancy-Related Low Back and Pelvic Pain." *Journal of Obstetric, Gynecologic & Neonatal Nursing* 46 (3): 334-346.

Ko, H., S. Le, and S. Kim. 2016. "Effects of Yoga on Dysmenorrhea: A Systematic Review of Randomized Controlled Trials." *Alternative and Integrative Medicine* 5 (4): 1-5.

Lee, M., J. Kim, J. Ha, K. Boddy, and E. Ernst. 2009. "Yoga for Menopausal Symptoms: A Systematic Review." *Menopause* 16 (3): 602-608.

Lincoln, G. 2002. "The Irritable Male Syndrome." *Reproduction, Fertility and Development* 13 (8): 567-576.

Marnach, M., K. Ramin, P. Ramsey, S. Song, J. Stensland, and K. An. 2003. "Characterization of the Relationship Between Joint Laxity and Maternal Hormones in Pregnancy." *Obstetrics and Gynecology* 101 (2): 331-335.

Matsumoto, T., T. Ushiroyama, T. Kimura, T. Hayashi, and T. Moritani. 2007. "Altered Autonomic Nervous System Activity as a Potential Etiological Factor of Premenstrual Syndrome and Premenstrual Dysphoric Disorder." *Biopsychosocial Medicine* 1: 24.

McGovern, C., and C. Cheung. 2018. "Yoga and Quality of Life in Women With Primary Dysmenorrhea: A Systematic Review." *Journal of Midwifery & Women's Health* 63:470-482.

Miner, S., S. Robins, Y. Zhu, K. Keeren, V. Gu, S. Read and P. Zelkowitz. 2018. "Evidence for the Use of Complementary and Alternative Medicines During Fertility Treatment: A Scoping Review." *BMC Complementary and Alternative Medicine* 18(1): 158.

Narendran, S., R. Nagarathna, V. Narendran, S. Gunasheela, and H. Nagendra. 2005. "Efficacy of Yoga on Pregnancy Outcome." *Journal of Alternative and Complementary Medicine* 11 (2): 237-244.

National Center for Health Statistics. 2019. "Infertility." Last modified December 2, 2019. www.cdc.gov/nchs/fastats/infertility.htm

Nygaard, I., C. Saltzman, M. Whitehouse, and F. Hankin. 1989. "Hand Problems in Pregnancy." *American Family Physician* 39:123-126.

Olive, D., and L. Schwartz. 1993. "Endometriosis." *New England Journal of Medicine* 328:1759-1769.

Oron, G., E. Allnutt, T. Lackman, T. Sokal-Arnon, H. Holzer, and J. Takefman. 2015. "A Prospective Study Using Hatha Yoga for Stress Reduction Among Women Waiting for IVF Treatment." *Reproductive Biomedicine Online* 30 (5): 542-548.

Quinlan, J., and D. Ashley Hill. 2003. "Nausea and Vomiting of Pregnancy." *American Family Physician* 68 (1): 121-128.

Riley, K., and E. Drake. 2013. "The Effects of Prenatal Yoga on Birth Outcomes: A Systematic Review of the Literature." *Journal of Prenatal & Perinatal Psychology & Health* 28 (1): 3-19.

Rosano, G., C. Vitale, G. Marazzi, and M. Volterrani. 2007. "Menopause and Cardiovascular Disease: The Evidence." *Climacteric* 10 (Suppl. 1): 19-24.

Sampson, J. 1927. "Peritoneal Endometriosis Due to the Menstrual Dissemination of Endometrial Tissue Into the Peritoneal Cavity." *American Journal of Obstetrics and Gynecology* 14:422-469.

Sasson, I., and H. Taylor. 2008. "Stem Cells and the Pathogenesis of Endometriosis." *Annals of the New York Academy of Sciences* 1127:106-115.

Sherman, S. 2005. "Defining the Menopausal Transition." *American Journal of Medicine* 118 (Suppl. 12): 3-7.

Stein, A., R. Pearson, S. Goodman, E. Rapa, A. Rahman, M. McCallum, L. Howard, and C. Pariante. 2014. "Effects of Perinatal Mental Disorders on the Fetus and Child." *Lancet* 384 (9956): 1800-1819.

U.S. Department of Health and Human Services. 2004. *Bone Health and Osteoporosis: A Report of the Surgeon General.* Rockville, MD: DHHS.

Wotring, V. 2012. *Space Pharmacology.* New York: Springer-Verlag.

Wu, W., O. Meijer, K. Uegaki, J. Mens, J. van Dieën, P. Wuisman, and H. Ostgaard. 2004. "Pregnancy-Related Pelvic Girdle Pain (PGP), I: Terminology, Clinical Presentation, and Prevalence." *European Spine Journal* 13:575-589.

Zaafrane, F., R. Faleh, W. Melki, M. Sakouhi, and L. Gaha. 2007. "An Overview of Premenstrual Syndrome." *Journal of Gynecology, Obstetrics and Biology of Reproduction* 36 (7): 642-652.

Chapter 8

Alirezaei, M., C.C. Kemball, C.T. Flynn, M.R. Wood, J.L. Whitton, and W.D. Kiosses. 2010. "Short-Term Fasting Induces Profound Neuronal Autophagy." *Autophagy* 6 (6): 702-710. https://doi.org/10.4161/auto.6.6.12376.

Brandhorst, S., I.Y. Choi, M. Wei, et al. 2015. "A Periodic Diet That Mimics Fasting Promotes Multi-System Regeneration, Enhanced Cognitive Performance, and Healthspan." *Cell Metabolism* 22 (1): 86-99. https://doi.org/10.1016/j.cmet.2015.05.012.

Bressa, C., M. Bailén-Andrino, J. Pérez-Santiago, et al. 2017. "Differences in Gut Microbiota Profile Between Women With Active Lifestyle and Sedentary Women." *PLOS ONE* 12 (2): e0171352. https://doi.org/10.1371/journal.pone.0171352 .

Caccialanza, R., E. Cereda, F. De Lorenzo, et al. 2018. "To Fast, or Not to Fast Before Chemotherapy, That Is the Question." *BMC Cancer* 18:337. https://doi.org/10.1186/s12885-018-4245-5.

Camilleri, M. 2009. "Serotonin in the Gastrointestinal Tract." *Current Opinion in Endocrinology, Diabetes, and Obesity* 16 (1): 53-59. https://doi.org/10.1097/med.0b013e32831e9c8e.

Clemente, J., E. Pehrsson, M. Blaser, K. Sandhu, et al. 2015. "The Microbiome of Uncontacted Amerindians." *Science Advances,* April 17: E1500183.

Dalen, J., B.W. Smith, B.M. Shelley, A.L. Sloan, L. Leahigh, and D. Begay. 2010. "Pilot Study: Mindful Eating and Living (MEAL): Weight, Eating Behavior, and Psychological Outcomes Associated With a Mindfulness-Based Intervention for People With Obesity." *Complementary Therapies in Medicine* 18 (6): 260-264. https://doi.org/10.1016/j.ctim.2010.09.008.

Drake, J.C., R.J. Wilson, and Z. Yan. 2016. "Molecular Mechanisms for Mitochondrial Adaptation to Exercise Training in Skeletal Muscle." *The FASEB Journal* 30 (1): 13-22. https://doi.org/10.1096/fj.15-276337.

Flynn, M.G., B.K. McFarlin, and M.M. Markofski. 2007. "The Anti-Inflammatory Actions of Exercise Training." *American Journal of Lifestyle Medicine* 1 (3): 220-235. https://doi.org/10.1177/1559827607300283.

Glick, D., S. Barth, and K.F. Macleod. 2010. "Autophagy: Cellular and Molecular Mechanisms." *Journal of Pathology* 221 (1): 3-12. https://doi.org/10.1002/path.2697.

Goodpaster, B.H., and L.M. Sparks. 2017. "Metabolic Flexibility in Health and Disease." *Cell Metabolism* 25 (5): 1027-1036. https://doi.org/10.1016/j.cmet.2017.04.015.

Grundmann, O., and S.L. Yoon. 2010. "Irritable Bowel Syndrome: Epidemiology, Diagnosis and Treatment: An Update for Health-Care Practitioners." *Journal of Gastroenterol Hepatology* 25 (4): 691-699. https://doi.org/10.1111/j.1440-1746.2009.06120.x.

Hagen, A. 2019. "It's The End of an Era for NYC Yoga." *Yoga Journal*. www.yogajournal.com/yoga-101/its-the-end-of-an-era-for-nyc-yoga/.

Helander, H.F., and L. Fändriks. 2014. "Surface Area of the Digestive Tract—Revisited." *Scandinavian Journal of Gastroenterology* 49 (6): 681-689.

Hoffman, H.J., E.K. Ishii and R.H. MacTurk. 1998. "Age-Related Changes in the Prevalence of Smell/Taste Problems Among the United States Adult Population. Results of the 1994 Disability Supplement to the National Health Interview Survey (NHIS)." *Annals of the New York Academy of Science* 855:716-722. https://doi.org/10.1111/j.1749-6632.1998.tb10650.x.

Hold, G.L., M. Smith, C. Grange, E.R. Watt, E.M. El-Omar, and I. Mukhopadhya. 2014. "Role of the Gut Microbiota in Inflammatory Bowel Disease Pathogenesis: What Have We Learnt in the Past 10 Years?" *World Journal of Gastroenterology* 20, no. 5 (February): 1192-1210.

Jaiswal, Y.S., and Williams, L.L. 2016. "A Glimpse of Ayurveda—The Forgotten History and Principles of Indian Traditional Medicine." *Journal of Traditional and Complementary Medicine* 7 (1): 50-53. https://doi.org/10.1016/j.jtcme.2016.02.002.

Jordan, C.H., W. Wang, L. Donatoni, and B.P. Meier. 2014. "Mindful Eating: Trait and State Mindfulness Predict Healthier Eating Behavior." *Personality and Individual Differences* 68:107-111.

Kaur, A. n.d. The Egg and the Orange ~ The Natural Wisdom of the Body. Kundalini Yoga Teachers' Association. https://www.kundaliniyoga.org.uk/the-egg-and-the-orange/

Karbowska J., and Z. Kochan. 2012. "Intermittent Fasting Up-Regulates Fsp27/Cidec Gene Expression in White Adipose Tissue." *Nutrition* 28:294-299. https://doi.org/10.1016/j.nut.2011.06.009.

Klein, A.V., and H. Kiat. 2015. "Detox Diets for Toxin Elimination and Weight Management: A Critical Review of the Evidence." *Journal of Human Nutrition and Dietetics* 28 (6): 675-686. https://doi.org/10.1111/jhn.12286.

Kristeller, J.L., and K.D. Jordan. 2018. "Mindful Eating: Connecting With the Wise Self, the Spiritual Self." *Frontiers in Psychology* 9 (August 14):1271. https://doi.org/10.3389/fpsyg.2018.01271.

Li, Y., and C. Owyang. 2003. "Musings on the Wanderer: What's New in Our Understanding of Vago-Vagal Reflexes? V. Remodeling of Vagus and Enteric Neural Circuitry After Vagal Injury." *American Journal of Physiology-Gastrointestinal and Liver Physiology* 285 (3): G461-G469.

Macy, D. 2017. "Eat Like a Yogi: A Yoga Diet Based in Ayurvedic Principles." *Yoga Journal*. Updated June 1, 2017. www.yogajournal.com/lifestyle/eat-like-a-yogi.

Matsumoto, M., R. Inoue, T. Tsukahara, K. Ushida, H. Chiji, N. Matsubara, and H. Hara. 2008. "Voluntary Running Exercise Alters Microbiota Composition and Increases n-Butyrate Concentration in the Rat Cecum." *Bioscience, Biotechnology, and Biochemistry* 72 (2): 572-576.

Mohr, A.E., R. Jäger, K.C. Carpenter, C.M. Kerksick, M. Purpura, J.R. Townsend, N.P. West, et al. 2020. "The Athletic Gut Microbiota." *Journal of the International Society of Sports Nutrition* 17: 1-33.

Muoio, D.M. 2014. "Metabolic Inflexibility: When Mitochondrial Indecision Leads to Metabolic Gridlock." *Cell* 159 (6): 1253-1262.

Ng, T.K.S., J. Fam, L. Feng, I.K-M. Cheah, C.T-Y. Tan, F. Nur, S.T. Wee, et al. 2020 "Mindfulness Improves Inflammatory Biomarker Levels in Older Adults With Mild Cognitive Impairment: A Randomized Controlled Trial." *Translational Psychiatry* 10 (1): 1-14.

Pietrocola, F., J.M. Bravo-San Pedro, L. Galluzzi, and G. Kroemer. 2017. "Autophagy in Natural and Therapy-Driven Anticancer Immunosurveillance." *Autophagy* 13 (12): 2163-2170. https://doi.org/10.1080/15548627.2017.1310356.

Pinto, J.M. 2011. "Olfaction." *Proceedings of the American Thoracic Society* 8 (1): 46-52. https://doi.org/10.1513/pats.201005-035RN.

Pullen, P.R., S.H. Nagamia, P.K. Mehta, et al. 2008. "Effects of Yoga on Inflammation and Exercise Capacity in Patients With Chronic Heart Failure." *Journal of Cardiac Failure* 14 (5): 407-413. https://doi.org/10.1016/j.cardfail.2007.12.007.

Pullen, P.R., W.R. Thompson, D. Benardot, et al. 2010. "Benefits of Yoga for African American Heart Failure Patients." *Medicine & Science in Sports & Exercise* 42 (4): 651-657. https://doi.org/10.1249/MSS.0b013e3181bf24c4.

Qin, H-Y., C-W. Cheng, X-D. Tang, and Z-X. Bian. 2014. "Impact of Psychological Stress on Irritable Bowel Syndrome." *World Journal of Gastroenterology* 20 (39): 14126.

Rioux, J., C. Thomson, and A. Howerter. 2014. "A Pilot Feasibility Study of Whole-Systems Ayurvedic Medicine and Yoga Therapy for Weight Loss." *Global Advances in Health and Medicine* 3 (1): 28-35. https://doi.org/10.7453/gahmj.2013.084.

Rizopoulos, N. 2017. "Get Twisted." *Yoga Journal.* Updated April 12, 2017. www.yogajournal.com/lifestyle/get-twisted.

Saltzman, J.A., S. Musaad, K.K. Bost, B.A. McBride, and B.H. Fiese. 2019. "Associations Between Father Availability, Mealtime Distractions and Routines, and Maternal Feeding Responsiveness: An Observational Study." *Journal of Family Psychology* 33 (4): 465.

Schnabel, L., E. Kesse-Guyot, B. Allès, et al. 2019. "Association Between Ultraprocessed Food Consumption and Risk of Mortality Among Middle-Aged Adults in France." *JAMA Internal Medicine* 179 (4): 490-498. https://doi.org/10.1001/jamainternmed.2018.7289.

Schumann, D., D. Anheyer, R. Lauche, G. Dobos, J. Langhorst, and H. Cramer. 2016. "Effect of Yoga in the Therapy of Irritable Bowel Syndrome: A Systematic Review." *Clinical Gastroenterology and Hepatology* 14 (12): 1720-1731. https://doi.org/10.1016/j.cgh.2016.04.026.

Sears, M., and S. Genuis. 2012. "Environmental Determinants of Chronic Disease and Medical Approaches: Recognition, Avoidance, Supportive Therapy, and Detoxification." *Journal of Environmental and Public Health*, Article ID 356798. https://doi.org/10.1155/2012/356798 .

Sears, S., and S. Kraus. 2009. "I Think Therefore I Om: Cognitive Distortions and Coping Style as Mediators for the Effects of Mindfulness Meditation on Anxiety, Positive and Negative Affect, and Hope." *Journal of Clinical Psychology* 65 (6): 561-573.

Sender, R., S. Fuchs, and R. Milo. 2016. "Revised Estimates for the Number of Human and Bacteria Cells in the Body." *PLOS Biology* 14 (8): e1002533. https://doi.org/10.1371/journal.pbio.1002533.

Shao, Y., S.C. Forster, E. Tsaliki, et al. 2019. "Stunted Microbiota and Opportunistic Pathogen Colonization in Caesarean-Section Birth." *Nature* 574:117-121. https://doi.org/10.1038/s41586-019-1560-1.

Sharma, S., S. Puri, T. Agarwal, and V. Sharma. 2009. "Diets Based on Ayurvedic Constitution—Potential for Weight Management." *Alternative Therapies in Health and Medicine* 15 (1): 44-47.

Sivananda Yoga Europe. n.d. "Vegetarianism." Accessed August 1, 2020. www.sivananda.eu/en/diet/vegetarianism.html.

Srour, B., L.K. Fezeu, E. Kesse-Guyot, et al. 2019. "Ultra-Processed Food Intake and Risk of Cardiovascular Disease: Prospective Cohort Study (NutriNet-Santé)." *BMJ* 365:l1451. https://doi.org/10.1136/bmj.l1451.

Tabas, I., and C.K. Glass. 2013. "Anti-Inflammatory Therapy in Chronic Disease: Challenges and Opportunities." *Science* 339 (6116): 166-172. https://doi.org/10.1126/science.1230720.

Torres, J., P. Ellul, J. Langhorst, A. Mikocka-Walus, M. Barreiro-de Acosta, C. Basnayake, N.J. Sheng Ding, D. Gilardi, K. Katsanos, G. Moser, R. Opheim, C. Palmela, G. Pellino, S. Van der Marel, and S.R. Vavricka. 2019. "European Crohn's and Colitis Organisation Topical Review on Complementary Medicine and Psychotherapy in Inflammatory Bowel Disease." *Journal of Crohn's and Colitis* 13, no. 6 (June): 673e-685e. https://doi.org/10.1093/ecco-jcc/jjz051.

Vennemann, M.M., T, Hummel, and K. Berger. 2008. "The Association Between Smoking and Smell and Taste Impairment in the General Population." *Journal of Neurology* 255 (8): 1121-1126. https://doi.org/10.1007/s00415-008-0807-9.

Yatsunenko, T., F.E. Rey, M.J. Manary, et al. 2012. "Human Gut Microbiome Viewed Across Age and Geography." *Nature* 486 (7402): 222-227. https://doi.org/10.1038/nature11053.

Younge, N., J.R. McCann, J. Ballard, C. Plunkett, S. Akhtar, F. Araújo-Pérez, A. Murtha, D. Brandon, and P.C. Seed. 2019. "Fetal Exposure to the Maternal Microbiota in Humans and Mice." *JCI Insight* 4 (19): HTTPS://DOI.ORG/10.1172/jci.insight.127806.

Yun, C.W., and S.H. Lee. 2018. "The Roles of Autophagy in Cancer." *International Journal of Molecular Science* 19 (11): 3466. https://doi.org/10.3390/ijms19113466.

Zeballos, E., and B. Restrepo. 2018. *Adult Eating and Health Patterns: Evidence From the 2014-16 Eating & Health Module of the American Time Use Survey.* No. 1476-2019-2776. https://www.ers.usda.gov/publications/pub-details/?pubid=90465.

Chapter 9

Birch, J. 2011. "The Meaning of Hatha in Early Hathayoga." *Journal of the American Oriental Society* 131 (4): 527-554.

Cramer, H., R. Lauche, J. Langhorst, and G. Dobos. 2016. "Is One Yoga Style Better Than Another? A Systematic Review of Associations of Yoga Style and Conclusions in Randomized Yoga Trials." *Complementary Therapies in Medicine* 25:178-187.

Mallinson, J. 2011. "Haṭha Yoga" In Vol. 3 of *The Brill Encyclopedia of Hinduism*. www.academia.edu/1317005/Ha%E1%B9%ADha_Yoga_entry_in_Vol_3_of_the_Brill_Encyclopedia_of_Hinduism.

Index

About the Authors

Andrew McGonigle, MD, has been practicing yoga and meditation for over 15 years and teaching yoga since 2009. He teaches anatomy and physiology in many yoga teacher training courses and leads his own international workshops.

Although McGonigle studied anatomy in great detail during medical school, he learned to look at it from a different angle and create ways to make it relevant to yoga. He enrolled in hands-on dissection classes focused on fascia and spent his spare time rereading anatomy books, listening to podcasts, and talking about anatomy with anyone who would listen.

McGonigle previously contributed a monthly article to *Om Yoga & Lifestyle* magazine ("360° Yoga With Doctor Yogi") and wrote two chapters for *Yoga Teaching Handbook: A Practical Guide for Yoga Teachers and Trainees.* He is also the author of *Supporting Yoga Students With Common Injuries and Conditions.* McGonigle resides in Los Angeles.

Matthew Huy (pronounced "hooey") has been teaching yoga since 2005. He teaches anatomy and physiology in many teacher training courses in addition to providing mentoring and professional development workshops to yoga teachers.

While studying biology in college, Huy discovered the joy of movement when he enrolled in a dance class and then a yoga class. A few years later, he changed tack and went on to complete a bachelor of arts degree in dance from California State University at Long Beach and completed teacher training courses in yoga, Pilates mat work, TRX, and Thai yoga massage. In 2021, he completed a master of science degree in sport, health, and exercise science at Brunel University at London, where he focused on exercise physiology and pain science. His master's research centered on the impact of yoga teachers' language on their students. Huy resides near London in the United Kingdom.

You read the book—now complete the companion CE exam to earn continuing education credit!

Find and purchase the companion CE exam here:
US.HumanKinetics.com/collections/CE-Exam
Canada.HumanKinetics.com/collections/CE-Exam

50% off the companion CE exam with this code

PY2023